Processing and Properties of Advanced Ceramics and Composites III

Processing and Properties of Advanced Ceramics and Composites III

Ceramic Transactions, Volume 225

Edited by
Narottam P. Bansal
Jitendra P. Singh
Jacques Lamon
Sung R. Choi

A John Wiley & Sons, Inc., Publication

Published by John Wiley & Sons, Inc., Hoboken, New Jersey.
Published simultaneously in Canada.

For general information on our other products and services or for technical support, please contact our Customer Care Department within the United States at (800) 762-2974, outside the United States at (317) 572-3993 or fax (317) 572-4002.

Wiley also publishes its books in a variety of electronic formats. Some content that appears in print may not be available in electronic formats. For more information about Wiley products, visit our web site at www.wiley.com.

Library of Congress Cataloging-in-Publication Data is available.

ISBN: 978-1-118-05998-2
ISSN: 1042-1122

oBook ISBN: 978-1-118-14444-2
ePDF ISBN: 978-1-118-14441-1

Printed in the United States of America.

10 9 8 7 6 5 4 3 2 1

Contents

TESTING, CHARACTERIZATION, AND MICROSTRUCTURE-PROPERTY RELATIONSHIPS

FUNCTIONALLY GRADED MATERIALS

CERAMIC PROCESSING

COMPOSITES PROCESSING

*Paper Presented at the 8th Pacific Rim Conference on Ceramic and Glass Technology, May 31-June 5, 2009

DIRECTIONAL SOLIDIFICATION AND MICROWAVE PROCESSING

Preface

Two international symposia "Innovative Processing and Synthesis of Ceramics, Glasses and Composites" and "Ceramic Matrix Composites" were held during Materials Science & Technology 2010 Conference & Exhibition (MS&T'10), Houston, TX, October 17-22, 2010. These symposia provided an international forum for scientists, engineers, and technologists to discuss and exchange state-of-the-art ideas, information, and technology on advanced methods and approaches for processing, synthesis and characterization of ceramics, glasses, and composites. A total of 91 papers were presented in the form of oral and poster presentations. Authors from 14 countries (Belgium, Brazil, Canada, China, Czech Republic, France, Germany, Japan, Mexico, Slovenia, South Korea, Spain, Turkey, and the United States) participated. The speakers represented universities, industries, and government research laboratories.

Seventeen papers on various aspects of synthesis, processing and properties of ceramics, glasses, and composites that were discussed at the symposia are included in this proceeding volume. Each manuscript was peer-reviewed using The American Ceramic Society review process.

The editors wish to extend their gratitude and appreciation to all the authors for their cooperation and contributions, to all the participants and session chairs for their time and effort, and to all the reviewers for their useful comments and suggestions. Financial support from The American Ceramic Society (ACerS) is gratefully acknowledged. Thanks are due to the staff of the meetings and publications departments of ACerS for their invaluable assistance.

It is our earnest hope that this volume will serve as a valuable reference for the researchers as well as the technologists interested in innovative approaches for synthesis and processing of ceramics and composites as well as their properties.

NAROTTAM P. BANSAL
J. P. SINGH
JACQUES LAMON
SUNG R. CHOI

Fiber Composites

INFLUENCE OF FIBER ORIENTATION ON THE MECHANICAL PROPERTIES AND MICROSTRUCTURE OF C/C-SIC COMPOSITE PLATES PRODUCED BY WET FILAMENT WINDING TECHNIQUE

Fabian Breede, Severin Hofmann, Enrico Klatt, Sandrine Denis
German Aerospace Center (DLR)
Institute of Structures and Design
Stuttgart, Germany

ABSTRACT

C/C-SiC composite plates were manufactured by wet filament winding and warm pressing. The liquid silicon infiltration (LSI) method was used to reach the ceramic state. Starting products were carbon fibres and a phenolic resin. The fibre orientation of the C/C-SiC composite plates was varied. The fibre orientations were set to ±15°, ±30°, ±45°, ±60° and ±75°, respectively. The main objective was to investigate the mechanical properties at room temperature depending on a change in fibre orientation. Furthermore the microstructures were studied. The tensile and flexural strength showed the expected dependency of fibre orientation in load direction. Finally the Young's modulus and Poisson's ratio were compared to analytical predictions for the different fibre orientations. For the analytical predictions the inverse laminate theory (ILT), as presented by Zebdi et al. (2008), was applied. The results from ILT and experiment show the same trend. The experimental procedure and first results are presented in this work.

INTRODUCTION

Ceramic matrix composites (CMC) are favorable as structural materials in aerospace engineering where high specific strengths and resistance to elevated temperatures are desirable[1,2]. In the last 10 years there has been a big effort in developing new types of carbon silicon carbide composites (C/C-SiC) that are suitable for future rocket propulsion applications[3-7]. Still the goal is to replace the currently used heavy superalloys in combustion chambers as well as in nozzle extensions of liquid fuel rocket engines. At the DLR the potential of the well known LSI processing route, also called reactive melt infiltration (RMI), originally using fibre preforms (e.g. fabrics) is investigated[8-12]. Now the challenge is to combine the wet filament winding technique with the LSI-route to provide a cost and time efficient process route which will generate suitable fibre reinforced ceramic materials for aerospace nozzle structures.

Today SiC-matrix composites can be processed by (1) a gas phase route, also referred to as chemical vapor infiltration (CVI), (2) a liquid phase route including the polymer impregnation/ Pyrolysis (PIP) and liquid silicon infiltration (LSI), as well as (3) a ceramic route, i.e. a technique combining the impregnation of the reinforcement with a slurry and a sintering step at high temperature and high pressure.

The objective of this paper is the investigation of mechanical properties and microstructures of C/C-SiC composite plates manufactured by the combination of wet filament winding technique and liquid silicon infiltration route (LSI).

3

EXPERIMENTAL PROCEDURE

Fabrication method

The C/C-SiC composite plates were fabricated by a three-step procedure. This process consists of 1st Carbon Fibre Reinforced Plastic (CFRP) green body generation by wet filament winding and warm pressing, 2nd build-up of a porous C/C preform and 3rd build-up of SiC matrix by melt infiltration of silicon given in Fig. 1.

1st step: In order to manufacture rotation symmetrical rocket engine structures such as nozzles, filament winding technique is necessary. In a wet filament cross winding process, continuous HTA 6k carbon fibre is impregnated during winding with a phenolic resin and wound onto a rotating mandrel in a predetermined pattern to form a interweaved regular laminate. Since the mechanical characterization is much easier to achieve on plate than tubular samples, in the next step the un-cured winding form is cut along the rotation-axis and then unrolled to a plate shape. The uncured composite preform is then cured by warm pressing. Spacers are used in order to achieve a predetermined fibre volume content of about 60 percent.

2nd step: The polymer based CFRP green body plates are pyrolysed in an inert gas atmosphere at temperatures above 900°C leading to a porous C/C preform.

3rd step: In the last process step, the porous, fibre reinforced preform (C/C) is infiltrated with molten silicon in vacuum. The molten silicon reacts with the carbon and forms the ceramic SiC matrix.

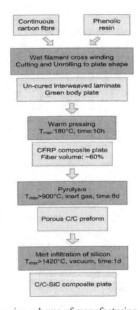

Figure 1. Processing scheme of manufacturing C/C-SiC plates.

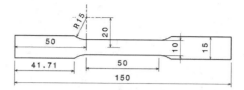

Figure 2. Dimensions of the tensile test samples in *mm*.

Material characterization

Density and open porosity of the C/C-SiC composite plates were measured by Archimedes method (acc. to DIN EN 993-1). Tensile strength of the C/C-SiC specimens was determined by using the tensile test at room temperature on an universal testing machine (Zwick Roell 1475) according to DIN EN 658-1. The signals of two strain gauges on each test sample were used to determine the Young's modulus and the elongation at fracture. The dimensions of the tensile test specimens are shown in Fig. 2.

The bending strength was measured by three-point-short-bending test at room temperature on an universal testing machine (Zwick Roell 1475) according to DIN EN 658-5. The dimensions of the bending test samples were 10mm (W) x 3mm (H) x 30mm (L). The relative position (fibre orientation) of the C/C-SiC test samples with respect to the loading direction is illustrated in Fig. 3. All mechanical properties are average values from 5-7 samples for each fibre orientation ($\pm\alpha$). The C/C-SiC samples were tested with a relative fibre orientation of $\pm15°$, $\pm30°$, $\pm45°$, $\pm60°$ and $\pm75°$.

The microstructure of the C/C-SiC composites was investigated by means of scanning electron microscopy (SEM, Zeiss Ultra 55) for each fibre orientation as mentioned above.

RESULTS

Density and open porosity

Open porosity and density are important material parameters which are determined after every process step. They are used for material characterization and serve as a quality assurance. Table 1 gives mean densities and open porosities of every process step of the C/C-SiC composite plates manufactured in this series. The properties relate to the manufacturing process depicted in Fig 1.

Mechanical properties of C/C-SiC composite plates

Table 1. Density and open porosity.

	CFRP	C/C	C/C-SiC
Density [g/cm³]	1.50 ± 0.03	1.38 ± 0.08	2.06 ± 0.08
Open porosity [vol.-%]	3.4 ± 1.9	15.2 ± 2.4	1.9 ± 0.4

Mechanical properties of C/C-SiC plates derived from tensile testing are depicted in Fig. 4. Typical stress-strain curves of each tested fibre orientation are shown in Fig. 5. As expected, tensile strength and Young's modulus increased with decreasing fibre orientations (α) with respect to the load direction. Maximum strain of failure was observed at a fibre orientation of $\pm45°$.

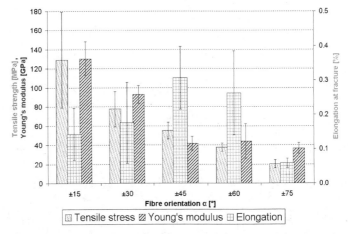

Figure 4. Tensile test properties of C/C-SiC composite samples with different fibre orientations.

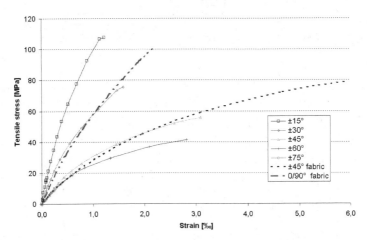

Figure 5. Typical stress-strain curves of C/C-SiC composite samples with different fibre orientations and 2 reference curves of C/C-SiC made of fabrics.

Bending strength behavior showed a similar characteristic. The results are depicted in Fig. 6.

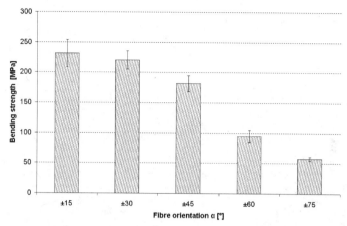

Figure 6. Bending strength (3pt. short beam bending test) for different fibre orientations.

Elastic properties calculated by Inverse Laminate Theory

The Young's modulus and Poisson's ratio obtained from tensile testing were compared to analytical predictions. The shear modulus was not considered in this study. The inverse laminate theory (ILT), as presented by Zebdi et al. (2008), was applied[13]. In the first step the properties of a virtual UD-ply were calculated from the elastic properties of the ±15°-composite (compare Table 2). E_{xx}, E_{yy} and v_{xy} were taken from tensile test data of the ±15°-composite. The Shear Modulus G_{xy} was estimated from Iosipescu test data from ±45° C/C-SiC. The properties of the generated virtual UD-ply were then used to calculate the elastic properties of the remaining composites with fibre orientations of ±30, ±45, ±60 and ±75°.

Table 2. Elastic Properties from tensile testing and Inverse Laminate Theory.

	±15°-Composite	Virtual UD-Ply
E_{xx} / GPa	141,0	170,2
E_{yy} / GPa	37,0	41,5
v_{xy} / -	0,6	0,3
G_{xy} / GPa	15,7	5,1

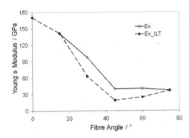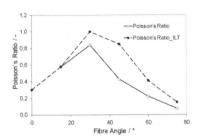

Figure 7. (a) Comparison of Young's Modulus and
(b) Poisson's ratio from tensile testing and ILT.

The results from ILT and experiment show the same trend in Young's modulus and Poisson's ratio (Fig. 7 (a) and (b)). Nevertheless the analytical solution differs from the experimental results in some extent. The largest relative differences were obtained at 45° angle. At this angle the calculated Young's modulus is only about 50% of the experimental modulus. In contrast the Poisson's ratio from ILT is almost double the experimental value at 45°. The main problems with ILT result from the sensitivity to small modifications of input data creating large variations of the solution. In addition the laminate theory naturally does not consider microstructural variations caused by the respective fibre orientations.

Microstructures
Typical microstructures of C/C-SiC composite plates are shown in Fig. 8 (a)-(b). It reveals the distribution of carbon fibres, C, SiC and Si in the composite (each phase has been denoted in the figure). The preferred segmentation of C/C-bundles is noticeable. However, the C/C-bundles show areas of SiC within the bundle. Furthermore, there is an inhomogeneous distribution of SiC rich C/C-bundles and SiC poor C/C-bundles across the component thickness.

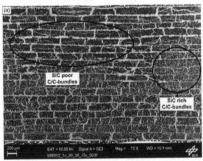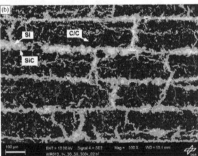

Figure 8. SEM micrographs of polished cross section surfaces of C/C SiC sample at α=±30°
magnification: (a) 72x and (b) 300x.

Fig. 9 shows micrographs comparing C/C-SiC composites manufactured via filament winding (left) and C/C-SiC made of fabrics (right). Both micrographs reveal similar C/C segmentation characteristics. The filament wound sample shows no undulations among the layers in contrast to the composite made of fabrics. The latter reveals less SiC in C/C-bundles than its counterpart.

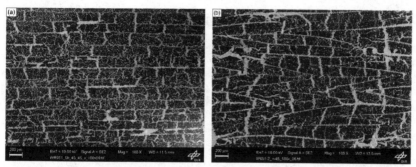

Figure 9. Comparison of SEM micrographs of C/C SiC cross section surfaces with fiber orientation α=±45° manufactured (a) by filament winding technique and (b) by stacking of

DISCUSSION

The micrographs of all fabricated C/C-SiC composite plates show C/C-bundle segmentation with an inhomogeneous distribution of SiC content across the thickness. Areas of SiC rich C/C-bundles weaken the composite due to the fact that more fibres are converted to SiC compared to C/C-bundles with only minor SiC content. Hence, the bonding of fibre and SiC-matrix is strong leading to brittle fracture behavior. C/C-SiC composites with a homogeneous distribution of dense C/C-bundles is favorable, where only the outer fibres of each C/C-bundle are converted to SiC. Wet winding process parameters like temperature of the resin, the amount of resin that is impregnated onto the fibre-bundle, winding speed and pre-curing of the CFRP preform (e.g. tempering) as well as the sizing of the C-fibres may have a decisive impact on the bonding between C-fibre and polymer matrix during pyrolysis. All plates were manufactured with the same process parameter set-up. Consequently non optimized process parameters lead to different types of microstructures. And these differences in microstructure have an essential influence on the mechanical behavior of C/C-SiC composites. In the future a process parameter study will eventually lead to an improved process parameter set up, which will yield to dense C/C-bundles after pyrolysis. This has to be investigated in more detail.

In general tensile strength and bending strength is mainly influenced by the fibre orientation, hence a fibre dominant dependency. Maximum strength values were reached when fibres were orientated in load direction ($\sigma_{tensile}$=130MPa, $\sigma_{bending}$=230MPa). The Young's modulus significantly increased when fibres were orientated in load direction (E=130GPa). The level of maximum strength is primarily predetermined by the microstructure and the fibre volume content in the ceramic state (here about 35-40 vol. %). Elongation at fracture had a maximum at a fibre orientation of ±45°. The reason is likely to be the combination of failure mechanisms, like failure due to tension and shearing of fibres.

CONCLUSION

For the first time C/C-SiC composite plates were successfully manufactured by filament winding of HTA-fibres and LSI route. The results show how fibre orientation influences the mechanical properties and give an overview on the microstructures. A future aim is to compare analytical and experimental results to provide a better understanding to improve the design of CMC composites as a structural material. Furthermore the results of this work will serve as reference data for upcoming material investigation on cylindrical specimens. In addition to the material characterization of C/C-SiC made of HTA-carbon fibres, the influence on mechanical properties of C/C-SiC using high modulus and high strength carbon fibres is currently investigated.

ACKNOWLEDGEMENT

Financial support has been provided by the German Research Council (Deutsche Forschungsgemeinschaft-DFG) in the framework of the Sonderforschungsbereich Trans-regio 40.

REFERENCES

[1]Jamet, J.F. and Lamicq, P.J. in: Naslain, R. (Eds.). High temperature Ceramic Matrix Composites. Woodhead, London, UK. p.735 (1993).

[2]Krenkel, W., Carbon fiber reinforced CMC for high performance structures. Int. J. Appl. Ceram. Technol, 1, 188–200 (2004).

[3]Haidn, O.J. et al., Development of Technologies for a CMC based Combustion Chamber. 2nd European Conf. for Aeros. Sciences (EUCASS) (2007).

[4]Schmidt, S. et al., Advanced ceramic matrix composite materials for current and future propulsion technology applications. Acta Astronautica, 55, 409-420 (2004).

[5]Beyer, S. et al., Advanced Composite Materials for Current and Future Propulsion and Industrial Applications. Advances in Science and Technology, Vol. 50, 174-181 (2006).

[6]Alting, J. et al., Hot-firing of an Advanced 40 kN Thrust Chamber. AIAA, 2001-3260 (2001).

[7]Naslain, R., Design, preparation and properties of non-oxide CMCs for application in engines and nuclear reactors: an overview. Composite Science and Technology, 64, 155–170 (2004).

[8]Hald, H., Weihs, H. and Reimer, T., Milestones towards hot CMC structures for operational space reentry vehicles. Proceedings of the 53rd International Astronautical Congress, Houston, USA (2002).

[9]Hillig, W.B., Making Ceramic Composites by Melt Infiltration. Amer. Ceram. Soc. Bulletin, 73(4), 56–62 (1994).

[10]Krenkel, W. and Kochendörfer, R., The LSI Process - A Cost Effective Processing Technique for Ceramic Matrix Composites. Proc. Int. Conf. on Adv. Mater (1996).

[11]Fabig, J. and Krenkel, W., Principles and New Aspects in LSI-Processing. 9th CIMTEC-World Ceramics Congress & Forum on New Materials (1998).

[12]Heidenreich, B. et al., C/C-SiC Materials with Unidirectional Reinforcement. Proc. 6th Int. Conf. on High Temp. Ceram. Matrix Comp, HT-CMC6 (2007).

[13]Zebdi, O., Boukhili, R. and Trochu, F., An Inverse Approach Based on Laminate Theory to Calculate the Mechanical Properties of Braided Composites, Journal of Reinforced Plastics and Composites, Online First, published on July 31, (2008).

HIGH-TEMPERATURE INTERLAMINAR TENSION TEST METHOD DEVELOPMENT FOR CERAMIC MATRIX COMPOSITES

Todd Z. Engel, Wayne S. Steffier, and Tony Magaldi
Hyper-Therm High-Temperature Composites, Inc.
Huntington Beach, CA, USA

ABSTRACT

Ceramic Matrix Composite (CMC) materials are an attractive design option for various high-temperature structural applications. However, 2D fabric-laminated CMCs typically exhibit low interlaminar tensile (ILT) strengths, and interply delamination is a concern for some targeted applications. Currently, standard test methods only address the characterization of interlaminar tensile strengths at ambient temperatures, which is problematic given that nearly all CMCs are slated for service in elevated temperature applications. This work addresses the development of a new test technique for the high-temperature measurement of CMC interlaminar tensile properties.

INTRODUCTION

Hot structures fabricated from ceramic composite (CMC) materials are an attractive design option for the specialized components of future military aerospace vehicles and propulsion systems because of the offered potential for increased operating temperatures, reduction of component weight, and increased survivability. CMC materials exhibit the high-temperature structural capabilities inherent to ceramic materials, but with significantly greater toughness and damage tolerance, and with lessened susceptibility to catastrophic component failure than traditional monolithic ceramic materials – all afforded by the presence of continuous fiber reinforcement. Some potential applications for CMC replacement of metallic components in aeroengine applications include flaps and seals, turbine blades and vanes. Most CMC laminates are fashioned from an assemblage of two-dimensional (2D) woven fabrics. As a result, the thru-thickness direction typically lacks fiber reinforcement and exhibits significantly lower strengths and toughness than the in-plane directions. For this reason, there is concern that thermostructural components fabricated from 2D fabric may have inadequate interlaminar tensile (ILT) strengths for some applications, thereby increasing their vulnerability to interply delamination when subjected to high thru-thickness thermal gradients, acoustic/high cycle fatigue, impact damage, free edge effects, and/or applied normal loads. For these reasons, it is important to fully characterize the ILT strengths of candidate CMC materials systems when they are being considered for use in high-temperature structural applications.

Currently, the ILT strength of CMCs is evaluated as per ASTM C1468[1] ("Standard Test Method for Transthickness Tensile Strength of Continuous Fiber-Reinforced Advanced Ceramics at Ambient Temperature"). This test method, commonly referred to as the flatwise tension (FWT) test, involves the use of square of circular planform test coupons machined from flat plate stock of representative CMC material. The specimens are typically machined from thin-gage plate material; for this reason it is not possible to directly grip the specimen for the application of the thru-thickness normal loading required to initiate ILT failure. Instead, loading blocks are adhesively bonded to the opposing faces of the specimen in order to facilitate tensile loading in the thru-thickness direction. The lack of availability of high-strength, high-temperature structural adhesives currently precludes the applicability of this methodology to

11

elevated temperature testing. A robust test method for performing ILT measurements at elevated temperatures is currently lacking and must be addressed to enable the serious consideration of CMC materials for insertion into high-temperature structural applications.

PROPOSED HIGH-TEMPERATURE ILT SPECIMEN CONFIGURATION

This work targets the aforementioned problem by introducing an alternative interlaminar tension specimen design that is conducive to testing at elevated temperatures. The body of the proposed specimen is machined from flat CMC plate stock. V-shaped notches are then machined into the thickness, and a state of interlaminar tension is induced between the specimen notches with wedge-type fixtures loaded in compression (Figure 1). Because this configuration only requires the application of compressive loading, it is readily adaptable to testing at high temperatures with the use of monolithic ceramic test fixturing.

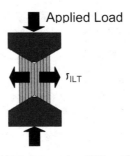

Figure 1. Proposed high-temperature ILT specimen.

The conceptual specimen geometry, illustrated in Figure 2, shows the four primary geometric parameters – namely the specimen thickness (t), the notch half-angle (φ), the notch spacing (b), and the specimen width (w).

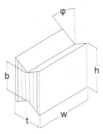

Figure 2. Proposed HT ILT specimen geometry.

The stresses induced in the loaded specimen were determined as a function of specimen geometry using a Strength of Materials analysis approach and the following key assumptions:

linear-elastic material behavior was assumed; the effects of the notch tip were ignored; small deformation behavior was assumed and the effects of friction between the specimen and the loading fixtures were ignored; load from the fixture is assumed to be applied uniformly to the inclined surface of the specimen. The components of stress induced in the specimen calculated from this analysis are derived as follows for an applied compressive load P, and are shown schematically in Figure 3.

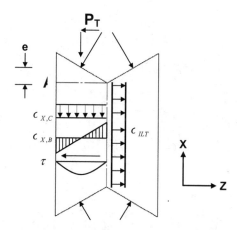

Figure 3. Induced components of stress in the proposed specimen.

From geometry, it can be shown that:

$$h = b + \frac{t}{TAN\varphi}$$

(1)

The resultant ILT force (P_T) applied to the specimen midplane is:

$$P_T = \frac{P}{2TAN\varphi}$$

(2)

Equation (2) indicates that large interlaminar tension forces can be generated in the test specimen with relatively small applied test loads if the notch angle is small. However, this may not be the best geometry for highly anisotropic materials. The Y-Z and X-Y planes through the notch apex B (planes AB and BC, respectively) are the critical sections for the ILT specimen. By virtue of symmetry, the specimen stress state is uniform along its' width (w). The normal compressive stress ($\sigma_{X,C}$) on the Y-Z plane AB due to the applied compression load is:

$$\sigma_{X,C} = -\frac{P}{wt}$$

(3)

Assuming a linear distribution, the maximum bending stresses ($\sigma_{X,B}$) on the Y-Z plane AB are:

$$\sigma_{X,B} = \pm\frac{6M}{w\left(\frac{t}{2}\right)^2} = \pm\frac{24M}{wt^2} \tag{4}$$

where the section bending moment is provided by:

$$M = eP_T \tag{5}$$

From Figure 2 geometry, it can be shown that:

$$e = \frac{t}{4TAN\varphi} \tag{6}$$

Upon substituting (5) into (4) yields:

$$\sigma_{X,B} = \pm\frac{3P}{wt(TAN\varphi)^2} \tag{7}$$

The maximum and minimum normal stresses on the Y-Z plane AB are found by superimposing the compressive and bending stresses, and are equal to:

$$c_{X,MAX/MIN} = c_{X,C} \pm c_{X,B} \tag{8}$$

The thru-thickness shear stresses on this plane AB are also of interest. The thru-thickness shear stress in flat plates has a parabolic distribution with zero stress on the surfaces and a maximum at the midplane. For the proposed ILT specimen, the maximum transverse shear stress occurs midway between points A and B and is equal to:

$$\tau_{MAX} = \frac{3P}{2wtTAN\varphi} \tag{9}$$

The interlaminar tensile stress at the specimen midplane ($\sigma_Z \equiv \sigma_{ILT}$) is equal to:

$$\sigma_{ILT} = \frac{2P_T}{wb} = \frac{P}{wbTAN\varphi} \tag{10}$$

As might be expected, the average ILT stress reduces to a function of the applied load (P), specimen notch half-angle (φ), and the ILT cross-sectional area of the test specimen (wb).

The closed form solution can be used to optimize the geometry for the proposed ILT specimen. For a pure ILT stress state at the specimen midplane, the in-plane bending stress at the apex (point B) needs to cancel the compressive stress from the applied load P. This condition can be found by setting equation (3) equal to equation (7). Upon substitution, the result is:

$$(TAN\varphi)^2 = 3 \quad \text{or} \quad \varphi = 60° \tag{11}$$

Therefore, the optimum specimen geometry for a pure ILT stress state at the specimen midplane occurs for a notch half-angle of 60°. For this geometry, the average ILT stress at the specimen midplane (σ_{ILT}) is:

$$\sigma_{ILT} = \frac{P}{wbTAN60°} = \frac{0.577P}{wb} \tag{12}$$

The laminate in-plane stresses on the Y-Z plane AB vary linearly from zero at the notch apex B to maximum compressive stress at point A equal to:

$$\sigma_{X,C-MAX} = -\frac{2P}{wt} \tag{13}$$

The maximum transverse shear stress occurs midway between points A and B on the Y-Z plane and is equal to:

$$\tau_{MAX} = \frac{0.866P}{wt} \tag{14}$$

Comparing equations (12) and (13), one notes that the maximum in-plane compressive stress, which occurs on the outer surfaces of the specimen, is 3.5 times higher than the midplane interlaminar tensile stress for b/t ratios of 1. An interlaminar tension failure mode should occur since the in-plane compressive strength of CMCs is generally one or two orders of magnitude higher than the interlaminar tensile strength. Dividing equation (9) by equation (10) yields the following:

$$\frac{\tau_{MAX}}{\sigma_{ILT}} = \frac{3b}{2t} \tag{15}$$

The material strength allowables can be substituted into this equation to generate a requirement for the ILT specimen geometric parameter b/t. Since equilibrium dictates that the interlaminar shear stress must equal the transverse shear stress at any point in the body, the interlaminar shear strength can be used as a lower bound for the transverse shear strength (as determined by Short Beam Shear test). Upon substitution:

$$\left(\frac{b}{t}\right) \leq \frac{2F_{ILS}}{3F_{ILT}} \tag{16}$$

Equation (16) defines the specimen geometry (for the 120° notch angle) as a function of the material's ILS (F_{ILS}) and ILT (F_{ILT}) strengths to ensure an interlaminar tension failure mode for a 120° specimen notch angle. This suggests that large b/t ratios may not be possible in that they may trigger an undesirable ILS failure.

PARAMETRIC SPECIMEN SIZING STUDY

A parametric study was performed to establish an optimized configuration for the proposed high-temperature ILT specimen. This geometry was determined by an experimental approach that identified the specimen geometry providing the most consistent measured correlation with ILT material strengths determined using conventional FWT test methods as per ASTM C1468. The specific geometric parameters that were targeted for investigation during the work were that of the notch half angle (φ) and notch spacing (b), as shown in Figure 2.

To empirically evaluate the test technique, a 6.4mm-thick silicon carbide fiber reinforced silicon carbide (SiC/SiC) CMC panel was manufactured using 2D CG Nicalon™ SiC fabric reinforcement (plain-weave) and chemical vapor infiltration (CVI) deposition techniques for the consolidation of a Pyrolitic Carbon (PyC) fiber coating and SiC ceramic matrix. The panel was machined into 25.4mm diameter cylindrical buttons to baseline the ILT properties of the CMC laminate using conventional FWT test methods. Additionally, six distinct geometric configurations of the proposed high-temperature ILT specimen were machined. The six variants were composed of two values of the notch half-angle (φ), namely 60° and 45°. This was supplemented with three values of the notch spacing parameter (b), namely 3.2mm, 6.4mm, and 12.7mm. For the 6.4mm-thick laminate, this equates to b/t ratios of ½, 1, and 2, respectively. The six variations of the two geometric parameters φ and b are shown schematically in Figure 4. Representative images of the machined specimens are shown in Figure 5.

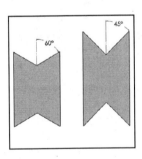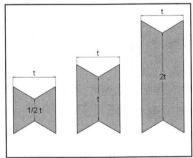

Figure 4. Specimen configurations of the parametric study.

Figure 5. Machined test specimens.

The flatwise tension samples were tested to provide a basis with which to compare the results of the parametric study. The cylindrical specimens were bonded to steel loading fixtures using a high-strength paste adhesive. The bonded assembly was then pulled in monotonic tension at a fixed displacement rate of 0.5mm per minute, inducing a interlaminar tensile failure between adjacent plies in the specimen. Six replicate FWT samples were tested, resulting in an average ILT strength of 9.2 MPa, with a standard deviation of 0.8 MPa. An image of the FWT test configuration is provided in Figure 6.

Stainless steel wedge fixtures were manufactured for the purpose of evaluating the room temperature ILT properties of the various proposed specimen configurations constituting the parametric study. The notched ILT specimens were inserted between the loading wedges and loaded in monotonic compression at a fixed displacement rate of 0.5mm per minute until failure was induced. An image of the room temperature test configuration used to evaluate the proposed ILT specimen is provided in Figure 6.

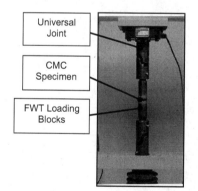

Figure 6. Room temperature test setups for conventional FWT specimens (left) and the proposed ILT specimen configuration (right).

The results of the parametric study showed the specimens machined with both the 60° and 45° notch half-angles exhibited similar trends with regard to measured ILT strengths. The samples with b/t ratios of ½ exhibited the highest average ILT strengths and highest levels of experimental scatter. The samples with b/t ratios of 1 exhibited comparatively lower average ILT strengths and reduced scatter; this configuration demonstrated good correlation with the FWT data. The samples with b/t ratios of 2 exhibited the lowest levels of scatter, but with ILT strengths significantly lower than those measured with FWT methods. The results of the parametric study are presented graphically in Figure 7, where measured ILT strengths are shown for both notch half-angles (φ) as a function of notch spacing (b). The mean ILT strength measured using the FWT method is also shown along with a plus/minus offset of two sample standard deviations for comparative purposes. The results of the parametric study are listed in Table 1. Both specimen configurations with and a b/t ratio of 1 demonstrated good correlation with the FWT results. However, the specimen with a notch half-angle (φ) of 60° was ultimately selected over that of the 45° specimen for further investigation due to the fact that it demonstrated lower levels of scatter for the limited data available. A mean ILT strength of 8.4 MPa with a standard deviation of 1.1 MPa was measured using this specimen configuration.

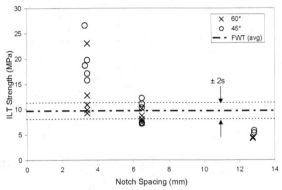

Figure 7. Parametric study test results.

Table I. Parametric Study Results

φ	b/t	No. Samples, n	Stressed Area (mm²)	ILT Strength Avg. (MPa)	ILT Strength Std. Dev. (MPa)	COV (%)
60°	½	5	1.60	13.2	5.7	43.2
	1	5	3.18	8.4	1.1	13.1
	2	2	6.35	4.3	0.1	2.3
45°	½	5	1.60	19.6	4.2	21.4
	1	5	3.18	9.6	2.3	24.0
	2	2	6.35	5.6	0.3	5.4
FWT		6	19.9	9.7	0.8	8.2

COMPREHENSIVE EVALUATION

Following the parametric evaluation and subsequent downselect of specimen geometry, additional test specimens were machined from the CMC plate for a more comprehensive evaluation of the chosen specimen configuration ($\varphi = 60°$, b/t = 1). This comprehensive evaluation included 17 additional tests performed at ambient conditions, and 7 tests performed at elevated temperature (1100°C). To facilitate the high-temperature testing, the test specimen was loaded with test fixtures machined from monolithic silicon carbide. The specimen was heated to the test temperature with a split tube furnace, and compressive load was transferred into the hot zone using alumina rods. The test procedure and loading rate was the same as with the room temperature testing. The high-temperature test setup is shown in Figure 8.

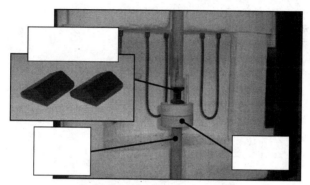

Figure 8. High-temperature test fixturing.

The testing of additional samples during the comprehensive evaluation resulted in a closer correlation of mean ILT strengths with the FWT results, but with an increase in experimental scatter. For the aggregated sum of the testing of the proposed specimen configuration evaluated under ambient conditions (22 total samples), the ILT strengths averaged 10.1 MPa with a standard deviation of 1.9 MPa. The 7 samples evaluated at elevated temperature (1100°C) measured a mean ILT strength of 11.2 MPa with a standard deviation of 2.9 MPa. Although this data is limited, initial indications appear to suggest that there is not a statistically significant difference between the ILT strengths measured at 25°C and 1100°C for the CG Nicalon™-based SiC/SiC material evaluated. This data is tabulated in Table 2 and presented graphically in Figure 9.

Table II. Comprehensive Test Results

φ	b/t	Temperature (°C)	No. of Tests, n	ILT Strength		
				Avg. (ksi) (MPa)	Std. Dev. (ksi) (MPa)	COV (%)
60°	1	25	22	10.1	1.9	18.9
60°	1	1100	7	11.2	2.9	25.9
FWT		25	6	9.7	0.8	8.2

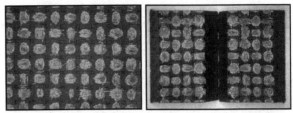

Flatwise Tension **Proposed ILT Method**

Figure 9. Comparison of fracture surfaces for the FWT and proposed ILT test methods.

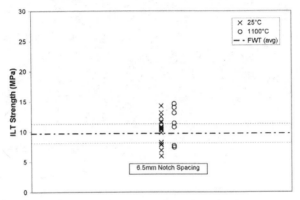

Figure 10. Comprehensive room temperature and elevated temperature results.

CONCLUSIONS

The proposed test technique for measuring the high-temperature ILT strength of CMC materials appears to demonstrate good correlation of mean ILT strengths to results obtained using more conventional FWT methodologies. From the parametric study of specimen geometry performed on 6.4mm-thick CMC material, the notch spacing to thickness (b/t) ratio of one produced the results most consistent with the FWT testing on the basis of comparing mean ILT strengths. Both notch half-angle (φ) values that were evaluated (60° and 45°) produced consistent results. Elevated temperature tests were performed at 1100°C, resulting in measured ILT strengths that were consistent with those measured at room temperature using both the proposed technique, and conventional FWT methods.

The distribution of ILT strengths measured from the 6.4mm laminate using the proposed test technique included higher levels of experimental scatter than was demonstrated with the conventional FWT technique. This is believed to be attributable to either the experimental method, or the relative sizing of the test specimen. Future work will examine different methods of loading the specimen and different notch-tip geometries. The stressed area of the proposed specimen geometry was 16 percent of that of the FWT specimen. This factor could potentially be influencing the distribution of ILT strength data if failures are originating from distributed volumetric or surface flaws within the test specimen, and thus subject to specimen size scaling effects. The relative sizing of the proposed high-temperature ILT specimen and the FWT specimen will also be investigated in future work.

ACKNOWLEDGEMENTS

We would like to acknowledge that this work has been funded by NAVAIR under contract number N68335-08-C-0491, and monitored by Dr. Sung Choi.

REFERENCES

[1]ASTM Standard C 1468-06: Standard Test Method for Transthickness Tensile Strength of Continuous Fiber-Reinforced Advanced Ceramics at Ambient Temperature, *Annual Book of ASTM Standards*, Vol. 15.01, ASTM International, West Conshohocken, PA, 2009.

LIFE LIMITING BEHAVIOR OF CERAMIC MATRIX COMPOSITES UNDER INTERLAMINAR SHEAR AT ELEVATED TEMPERATURES

Sung R. Choi[†]
Naval Air Systems Command
Patuxent River, MD 20670

ABSTRACT

A series of work to assess life limiting behavior of various ceramic matrix composites (CMCs) in interlaminar shear has been conducted at elevated temperatures using a double notch shear configuration. Similar to the case of in-plane tensile loading, most CMCs investigated exhibited life limiting behavior in interlaminar shear with their degree of life susceptibility depending on materials. The phenomenological life prediction model in interlaminar shear has been proposed and verified through experimental data determined from both stress rupture and constants stress-rate testing. The results obtained to date indicate that the power-law crack growth could be used as a means of a phenomenological life prediction methodology in CMCs subjected to interlaminar shear at elevated temperatures.

1. INTRODUCTION

The successful development and design of continuous fiber-reinforced ceramic matrix composites (CMCs) are dependent on thorough understanding of their mechanical, thermal, and chemical properties and their responses to related environments. Particularly, accurate evaluation of life limiting behavior under specified loading/temperature/environmental conditions is a prerequisite to ensure accurate life prediction of structural composite components.

Although CMCs have shown improved fracture resistance and increased damage tolerance compared with monolithic ceramic counterparts, inherent material/processing defects or cracks in the matrix-rich interlaminar and/or interface regions can still cause delamination under interlaminar tensile or shear stress, resulting in loss of stiffness or in some cases structural failure. Interlaminar tensile and shear strength behaviors of CMCs have been characterized in view of their unique architectures and importance in structural applications [1-4]. Because of the materials' inherent nature, CMCs are expected to be highly susceptible to life limiting even in interlaminar shear or tension particularly at elevated temperatures, resulting in strength degradation or shortened lives.

A series of work to determine life limiting behavior of CMCs in interlaminar shear at elevated temperatures has been performed using stress rupture and constant stress-rate tests by the author et al. [5-8] and by others [9]. A total of seven different CMCs with various architectures and material constituents have been used to date. This paper summarizes all the previous results and discusses life limiting behavior of the CMCs using a power-law type of phenomenological life model to quantify their life susceptibility in interlaminar shear.

2. EXPERIMENTAL PROCEDURES

2.1. Materials

Seven different continuous fiber-reinforced CMCs - three SiC fiber-reinforced SiC matrix composites, three SiC fiber-reinforced glass-ceramic matrix composites, one N720 fiber-reinforced

---

[†] Email address: sung.choi1@navy.mil

Al_2O_3 matrix composite - have been used for life prediction testing in interlaminar shear. Table 1 summarizes information regarding preforms and resulting laminates of the CMCs.

The 2-D woven, 5 harness-satin, Sylramic cloth preforms in the Sylramic SiC/SiC CMC were stacked and chemically vapor infiltrated with a thin BN-based interface coating followed by SiC matrix over-coating. Remaining matrix porosity was filled with SiC particulates and then with molten silicon at $1400°C$, a process termed slurry casting and melt infiltration [10]. Similar process was utilized in the Hi-Nic SiC/SiC CMC. The 2-D plain-woven SiC/SiC composite ('90 vintage) was processed through chemical vapor infiltration (CVI) into the fiber preforms. The cross-plied SiC/CAS-II and SiC/MAS-5 CMCs were fabricated through hot pressing followed by ceraming of the composites by a thermal process [11].

The 1-D Hi-Nic SiC/BSAS was manufactured via BN/SiC fiber interface coating, matrix slurry prepreg, tape laying-up, and then hot pressing [5]. The 2-D woven, 8 harness-satin $N720/Al_2O_3$ oxide/oxide CMC ('08 vintage) was fabricated via slurry-infiltration followed by consolidation and sintering. Significant porosity (about 25%) and microcracks in the matrix were typified of the oxide/oxide composite, a characteristic of this class of oxide/oxide CMCs for increased damage tolerance. No interface fiber coating was applied [12]. More detailed descriptions of the CMCs used in the work can be found elsewhere [5-8,10-12].

Table 1. Continuous fiber-reinforced CMCs used

No	Materials	Architecture	Fiber	Fiber volume fraction	Process & Laminates[†]	Manufacturer
1	MI SiC/SiC [6]	2-D woven	Sylramic SiC	0.33	iBN;SC;MI;5HS;0/90;8 plies; t=2.2mm;20epi	GEPSC[‡]
2	MI SiC/SiC [8]	2-D woven	Hi-Nic SiC	0.39	iBN;SC;MI;5HS;0/90;8 plies; t=2.2mm;20epi	GEPSC[‡]
3	SiC/SiC ('90) [6]	2-D woven	Nicalon SiC	0.39	CVI;plain;12 plies; t=3.5mm	E. I. Du Pont[‡]
4	SiC/CAS-II [6]	2-D c-plied	Nicalon SiC	0.39	HP; 18plies; t=3mm	Corning
5	SiC/MAS-5 [6]	2-D c-plied	Nicalon SiC	0.39	HP; 16 plies; t=3mm	Corning
6	SiC/BSAS [5]	1-D plied	Hi-Nic SiC	0.42	BN-SiC coat; slurry-prepreg; HP; 20 plies; t=4mm	NASA GRC
7	N720/Al₂O₃ [12] (oxide/oxide)	2-D woven	N720	0.45	Slurry;12 plies; t=2.7mm	ATK/COIC

[‡] Currently GE Ceramic Composite Products (CCP), Newark, DL
[†] iBN: *in situ* BN coated; SC: slurry casting; MI: melt infiltration; HS: harness satin; t: thickness; epi: ends per inch; CVI: chemical vapor infiltration; HP: hot pressing.

2.2. Life Prediction Testing in Interlaminar Shear at Elevated Temperatures

The double-notch-shear (DNS) test specimens were cut from each composite laminate. Typically, test specimens were 13-15 mm wide (W), 30 mm long (L), and in as-received laminate thickness (t). Two notches, 0.3 mm wide (h) and 6 mm (L_n) away from each other, were machined such that they were located in the middle of the specimen within ±0.05 mm so that interlaminar shear failure occurred along the plane between the notch ends, see Fig. 1.

Stress rupture testing: Stress rupture testing in interlaminar shear was conducted at elevated temperatures in air with DNS test specimens of the composites, where time to failure was determined as a function of applied shear stress. The test setup consisting of SiC fixtures is also shown in Fig. 1. In case of thinner composite panels, ceramic anti-buckling guides were used. Each test specimen was heated at a heating rate of 10 °C/min and held for about 20 min at test temperature for thermal equilibration prior to testing. Testing was conducted using either electromechanical (Instron Model 8562) or servohydraulic (Instron Model 8501) test frame. Test temperatures ranging from 1100 to 1316 °C and numbers of applied stresses and test specimens were dependent on materials and will be shown in the results section. Test procedure, in general, was followed in accordance with ASTM test method C 1425 [13]. The nominal applied interlaminar shear stress was calculated using the following relation

$$\tau = \frac{P}{WL_n} \tag{1}$$

where τ is the applied shear stress, P is the applied force, and W and L_n are the specimen width and the distance between the two notches, respectively (see Fig. 1).

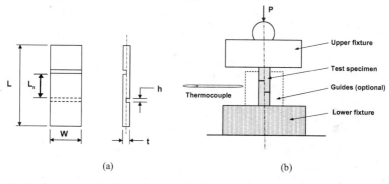

(a)   (b)

Figure 1. Configurations of double-notch shear (DNS) test specimen (a) and a schematic showing high-temperature test setup (b).

Constant shear stress-rate testing: Additional elevated-temperature testing was also conducted in air using DNS test specimens by employing constant shear stress-rate testing, where interlaminar shear strength was determined as a function of applied interlaminar shear stress rate. This type of testing is often called 'dynamic fatigue' when applied to brittle monolithic materials (such as glasses, glass-ceramics, advanced monolithic ceramics) in tension to evaluate their slow crack growth (SCG) behavior at ambient or elevated temperatures [14,15]. The degree of strength degradation with respect to applied test rate is indicative of the susceptibility to life limiting or SCG. The test frame, test specimen configuration, test setup, and heating-cooling mode utilized in constant shear stress-rate testing were identical to those used in the shear stress rupture testing. Test temperatures, numbers of applied stress and test specimens were dependent on materials and will be shown in the results section. One of the reasons for this additional testing was to assess the proposed life prediction model by predicting related life from one loading condition to another.

3. EXPERIMENTAL RESULTS AND DISCUSSION

3.1. Interlaminar Shear and Tensile Strengths of Various CMCs at Ambient Temperature

Figure 2 shows a database on ambient-temperature interlaminar shear and tensile strengths from the literature and determined by the author et al. [16] for a total of 15 different CMCs. The database shows significantly poor interlaminar strengths in which interlaminar shear and tensile strengths were about 1/8 and 1/20 of in-plane strength, respectively. The interlaminar shear strength was always greater than the interlaminar tensile counterpart. A relationship between the two strengths can be approximated from the figure as follows:

$$\tau_f \approx 3c_t \tag{2}$$

where τ_f and σ_t are interlaminar shear and tensile strengths, respectively. The relationship is rather surprising in view of no physical basis for the outcome. Note that the CMCs used differ significantly from each other in terms of architecture (woven or cross-plied, plain weave or multiple harness-satin), type of matrix (ceramic or glass ceramic), type of fiber (various SiCs or oxide), fiber/matrix interface properties, and type of processing (CVI, slurry casting, melt infiltration, or hot pressing). A similar trend in interlaminar strengths was also observed in crack growth resistances of G_I and G_{II} in interlaminar directions, although their relationship between G_I and G_{II} could not be quantified like Eq. (2), attributed to rising R-curve configurations [17]. There have been attempts to improve interlaminar strength properties by means of 3-D orthogonal, angle-interlocking weaving or matrix reinforcing, etc. However, pros and cons of those approaches still exist regarding issues on in-plane properties degradation, cost, affordability, and reliability.

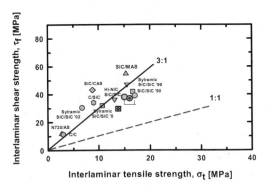

Figure 2. Interlaminar shear strengths vs. interlaminar tensile strengths for various continuous fiber-reinforced CMCs. An approximate relation between the two strengths (a 3:1 relation) is shown as a solid line in the figure [16].

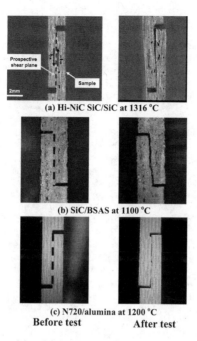

(a) Hi-NiC SiC/SiC at 1316 °C

(b) SiC/BSAS at 1100 °C

(c) N720/alumina at 1200 °C
Before test   **After test**

Figure 3. Typical examples of interlaminar shear failure at elevated temperatures for (a) Hi-NiC SiC/SiC [8], (b) SiC/BSAS [5], and (c) N720/alumina oxide/oxide CMCs. Side views of each DNS specimen before and after test.

3.2. Stress Rupture in Interlaminar Shear

Most of the DNS specimens tested in stress rupture failed in shear along their respective interlaminar shear planes. Some examples of shear failure for different CMCs are shown in Fig. 3. The results of stress rupture testing for the CMCs are summarized in Fig. 4, where by convention applied shear stress was plotted as a function of time to failure. The data show an evidence of life limiting behavior, where time to failure decreased with increasing applied shear stress. The best fit line was included for each CMC, based on the log (*time to failure*) vs. log (*applied interlaminar shear stress*) relation, which will be discussed later. A relatively large scatter in time to failure was observed in the Hi-Nic SiC/SiC CMC, similar to the feature observed in many CMCs and advanced monolithic ceramics subjected to stress rupture testing in tension or flexure at elevated temperatures

In general, the specimens tested at lower stresses showed somewhat smoother and cleaner fracture surfaces with less breakage of fiber tows than those tested at higher stresses. This implies that the matrix-rich interfacial failure through slow crack growth or damage accumulation might have been more predominant at lower stresses than at higher stresses. The time for slow crack growth along the fiber-matrix interfaces was not sufficient at higher stresses, resulting in a phenomenon reminiscent of 'fast fracture' that accompanies rough fracture surfaces with increased fiber tow damage.

Fracture surfaces of DNS tested specimens showed matrix-rich interface shear failure within one or two adjacent plies. Imprints of fibers were usually seen on the mating matrices. At higher applied stresses, some breakage of horizontal (warp) and vertical (weft or fill) fiber tows was observed. There were some 'blind' regions where no matrix or silicon was infiltrated between plies or tows, resulting in significant gaps or voids or pores in the composite. This certainly contributes to decrease in interlaminar shear properties of the composite. The common features of fracture surfaces aforementioned are presented in Fig. 5. However, it should be emphasized that the regions of slow crack growth, commonly discernable and demarcated in many monolithic brittle materials, did not appear in CMCs, which is a daunting challenge in CMCs as far as fractography is concerned.

3.3. Constant Stress-Rate Tests in Interlaminar Shear

All the specimens tested in constant stress-rate loading failed in interlaminar shear, similar to those in stress rupture. The results of constant stress-rate testing are shown in Fig. 6, where interlaminar failure stress (or interlaminar shear strength) was plotted against applied shear stress rate in a log-log scheme. The solid lines represent the best fit. In many cases, the overall interlaminar shear strength decreased with decreasing applied shear stress rates with the degree of strength degradation being dependent on materials. This aforementioned phenomenon of strength degradation with decreasing test rate, called 'dynamic fatigue' when referred to monolithic brittle materials in tension or in flexure [14,15], is an evidence of slow crack growth (SCG) occurring at the fiber-matrix interfaces along respective interlaminar shear plane. The greater degradation in strength means the greater susceptibility to SCG. Hence, the plain-woven SiC/SiC and N720/alumina CMCs showed increased resistance to SCG. The SCG parameters, an index of SCG susceptibility, will be discussed and determined for the CMCs in the next section. Based on both of the results in stress rupture and constant shear stress rate testing, it can be stated that life limiting behavior of the composites took place in interlaminar shear, either in static loading (stress rupture) or in time-varying loading (constant stress rate) condition.

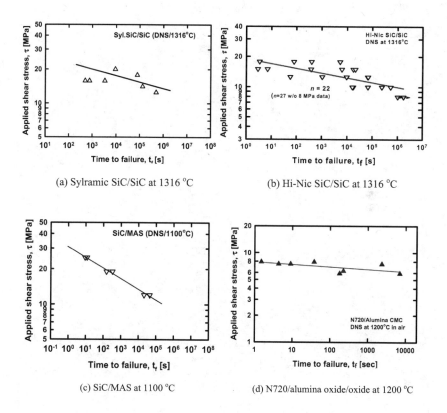

(a) Sylramic SiC/SiC at 1316 °C

(b) Hi-Nic SiC/SiC at 1316 °C

(c) SiC/MAS at 1100 °C

(d) N720/alumina oxide/oxide at 1200 °C

Figure 4. Results of shear stress rupture testing for (a) Sylramic SiC/SiC at 1316 °C [6], (b) Hi-Nic SiC/SiC at 1316 °C [8], (c) SiC/MAS at 1100 °C [5], and (d) N720/alumina oxide/oxide at 1200 °C. The best fit lines are included.

(a) Sylramic SiC/SiC   (b) Hi-Nic SiC/SiC

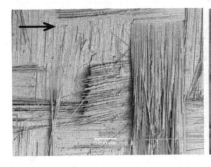

(c) N720/alumina oxide/oxide   (b) Hi-Nic SiC/SiC

Figure 5. (a) Overall fracture surface of Sylramic SiC/SiC, (b) Hi-Nic SiC/SiC showing fiber imprints in matrix-rich region, (c) N720/alumina oxide/oxide with fiber-tows breakage, and (d) Hi-Nic SiC/SiC showing unfilled blind region. The big arrows indicate the direction of shear loading.

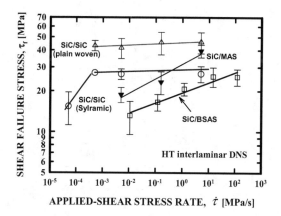

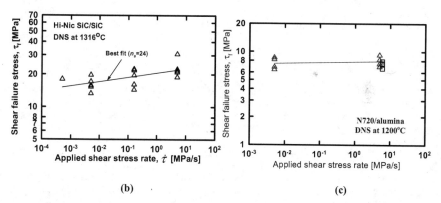

(b)   (c)

Figure 6. Results of constant shear stress rate testing for (a) Sylramic SiC/SiC (1316 °C), SiC/SiC (plain woven, 1200 °C)), SiC/CAS (1100 °C), and SiC/BSAS (1100 °C) [6], (b) Hi-Nic SiC/SiC (1316 °C) [8], and (c) N720/alumina oxide/oxide (1200 °C). The solid lines present the best fit.

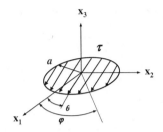

Figure 7. A circular crack residing in interlaminar matrix-rich region on the X_1-X_2 plane, subjected to shear stress (τ) [18].

4. DISCUSSION

4.1 Proposed Slow-Crack-Growth Model in Interlaminar Shear

A phenomenological slow crack growth (SCG) model to explain the life limiting behavior under interlaminar shear in both stress rupture and constant shear stress rate loading has been proposed and used for validation in some of CMCs [5-8]. This section describes briefly the underlying mechanics of the proposed model. The SCG model in mode II is similar in expression to the power-law relation in mode I, and takes a following, empirical formulation [5-8]:

$$v_s = \frac{da}{dt} = \alpha_s (K_{II} / K_{IIc})^{n_s}$$ (3)

where v_s, a, t, K_{II}, and K_{IIc} are crack-growth rate in interlaminar shear, crack size, time, mode II stress intensity factor, and mode II fracture toughness, respectively. α_s and n_s are life limiting (or SCG) parameters in interlaminar shear. The above formulation was fracture-mechanics based, in which a crack residing at matrix-rich interfacial regions is assumed to grow subcritically, eventually leading to an instability condition. It may be arguable as to whether a term 'crack' or fracture mechanics concept can be used for CMCs since fracture origin and subsequent crack growth region are not typically observable from fracture surfaces, attributed to their architectural complexity.

The generalized expression of K_{II} along the crack front of a penny- or half-penny shaped crack/flaw subjected to shear loading either on crack planes or on remote material body can take the following form [18]:

$$K_{II} = Y_s \tau a^{1/2} f(\theta, \varphi)$$ (4)

where Y_s is a crack geometry factor related to a function of $f(\theta, \varphi)$ with the angles θ and φ being related to load and a particular point of the crack front, as shown in Fig. 7. τ is the applied interlaminar shear stress. Using Eqs. (3) and (4) together with some mathematical manipulations, one can obtain time to failure (t_f) as a function of applied shear stress, as done for brittle materials in mode I loading [5-8]:

$$t_f = D_s [\tau]^{-n_s}$$ (5)

where

$$D_s = B_s\, [\tau_i]^{n_s - 2} \tag{6}$$

where $B_s = 2K_{IIc}[\alpha_s Y_s^2 (n_s - 2)]$ with Y_s being crack geometry factor, τ_i is the inert shear strength that is defined as a strength with no SCG, and the geometry function is simplified as $f(\theta, \varphi) = 1$ in the case of double-notch shear loading for a semi-infinite material body. Equation (5) can be expressed in a more convenient form by taking logarithms of both sides

$$\log t_f = -\, n_s \log \tau - \log D_s \tag{7}$$

which is identical in form to the case in mode I loading used in monolithic ceramics [14,15]. Life limiting parameters n_s and D_s in interlaminar shear can be determined based on Eq. (7), respectively, from the slope and the intercept of a linear regression analysis of the log (*individual applied interlaminar shear stress with units of MPa*) vs. log (*individual time to failure with unit of second*) data, Fig. 4.

In the case of constant shear stress-rate testing, applied shear stress to the test specimen at a given time, t, is expressed as follows:

$$\tau = \int_0^t \dot{\tau}\, dt = \dot{\tau}\, t \tag{8}$$

Similarly, using Eqs. (3), (4), and (8), interlaminar shear strength (τ_f) as a function of applied shear stress rate ($\dot{\tau}$) can be derived as follows [5-8]:

$$\tau_f = D_d\, [\dot{\tau}]^{\frac{1}{n_s + 1}} \tag{9}$$

where

$$D_d = [D_s (n_s + 1)]^{\frac{1}{n_s + 1}} \tag{10}$$

Taking logarithms of both sides of Eq. (9) yields

$$\log \tau_f = \frac{1}{n_s + 1} \log \dot{\tau} + \log D_d \tag{11}$$

which is identical in form to the case in mode I loading used in monolithic ceramics [14,15]. Life limiting parameters n_s and D_s in interlaminar shear can be determined based on Eq. (11), respectively, from the slope and the intercept of a linear regression analysis of the log (*individual applied interlaminar shear strength with units of MPa*) vs. log (*individual shear stress rate with unit of second*) data, Fig. 6.

Table 2. Summary of SCG parameters evaluated from the constant stress-rate test data (Fig. 6)

CMCs	Temp (°C)	SCG parameters	
		n_s	D_d
SiC/SiC (Sylramic)	1316	>90 (3*)	27
SiC/SiC (Hi-Nic)	1316	24	22
SiC/SiC (Nicalon)	1200	90	45
SiC/MAS	1100	8	32
SiC/BSAS	1100	11	19
N720/alumina	1200	>70	8

Note:

* indicate SCG parameter determined at shear stress rates $\dot{\tau} < 10^{-3}$ MPa

It is important to note that Eq. (10) indicates that for a given material/environmental condition the data on constant shear stress-rate tests can be converted to stress rupture data. In other words, the constant shear stress-rate data (n_s and D_d) can be used to predict stress rupture behavior and vice versa. Of course, the dominant failure mechanism should be identical for a given material/environment between the two loading configurations of stress rupture and constant stress rate.

4.2 Applications of the Model to Experimental Data: Verification

The proposed SCG formulation, Eq. (3), indicates that for a given material/environmental condition, crack velocity depends on K_{II} so that in principle SCG parameters (n_s and D_s or D_d; called life limiting or life prediction parameters, too) can be determined in any loading configuration which is either static, cyclic, or any time varying. Therefore, it is feasible to make a life prediction from one loading configuration to another provided that the same failure mechanism is operative, as aforementioned. In this section, the SCG parameters that were determined from the constant stress rate testing will be used to predict the stress rupture behavior and to validate the proposed SCG model.

The SCG parameters estimated based on the data in Fig. 6 and Eq. (11) are summarized in Table 2. At this point, it is helpful to categorize the degree of SCG based on the magnitude of the major SCG parameter n_s, as done in the mode I case. It has been generally classified for brittle materials under mode I that the SCG susceptibility (in the expression of $v = da/dt = \alpha[K_I/K_{Ic}]^n$) is significant for $n < 30$, intermediate for $n \sim 30$-40, and insignificant for $n \geq 50$ [14,15]. Hence, the Nicalon (plain-woven) SiC/SiC, the Sylramic SiC/SiCs at $\dot{\tau} > 10^{-3}$ MPa/s , and the N720/alumina oxide/oxide exhibited $n_s > 70$, indicative of their strong resistance to SCG, while the Hi-Nic SiC/SiC, SiC/MAS, SiC/BSAS, and Sylramic SiC/SiC at $\dot{\tau} < 10^{-3}$ MPa/s exhibited significant susceptibility to SCG with $n_s < 30$. However, it should be careful that the use of the categorization was not based on any physical correlation between n and n_s. More database is needed accordingly.

Using the SCG parameters already determined in Table 2, life prediction to stress rupture can be made based on Eqs. (5) and (10). The resulting prediction is presented as a solid line in Fig. 8. Except for the Sylramic SiC/SiC CMC, the prediction was in good agreement with the

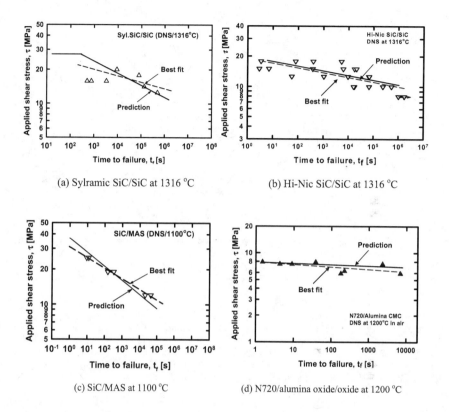

(a) Sylramic SiC/SiC at 1316 °C   (b) Hi-Nic SiC/SiC at 1316 °C

(c) SiC/MAS at 1100 °C   (d) N720/alumina oxide/oxide at 1200 °C

Figure 8. Comparison between the prediction and the stress rupture data for (a) Sylramic SiC/SiC at 1316 °C [6], (b) Hi-Nic SiC/SiC at 1316 °C [8], (c) SiC/MAS at 1100 °C [5], and (d) N720/alumina oxide/oxide at 1200 °C. The best fit and prediction lines are included.

best fit of other CMCs. Particularly for the Hi-Nic SiC/SiC composite, the SCG parameter of n_s=24 from the constant stress-rate loading was in very good agreement with n_s=22 determined from the stress rupture data based on Eq. (7). The D_s between the prediction and the data was also in good agreement. The good agreement indicates that the *overall* governing failure mechanism of the CMCs subjected to interlaminar shear might have remained almost unchanged, regardless of loading configurations and that the failure mechanism could be described by the power-law type of SCG formulation, Eq. (3). Of course, more data are needed for better prediction especially in the Sylramic SiC/SiC composite.

5. CONCLUDING REMARKS

The results of the work presented here indicate that the governing failure mechanism would not differ appreciably either in stress rupture or in constant stress-rate loading and that the overall failure mechanism could be described by the power-law type of SCG formulation, Eq. (3). The results also indicate that stress rupture or constant stress-rate testing can be utilized as a means of life-prediction test methodology in interlaminar shear for CMCs. Of course, stress rupture testing gives more realistic life data but it gives increased data scatter so that more test specimens are required, which is time consuming particularly at lower applied stresses. By contrast, constant stress-rate testing gives merits in simplicity, test economy (short test times), and less data scatter thus allowing less number of test specimens. However, caution should be stressed when more than one failure mechanism occur, associated with SCG, creep, and/or environments. In this case, use of constant stress-rate testing should be limited for the case of a short period of lives.

Exploration of detailed failure mechanism(s) associated in interlaminar shear at elevated temperatures was beyond the scope of this work. Microscopic/macroscopic aspects such as matrix/fiber interaction, matrix and fiber cracking, localized SCG, creep, and environments [19-21] need to be considered. It should be noted that the phenomenological model proposed in this work may incorporate other operative models such as viscous sliding, void nucleation, growth and coalescence, etc., which can be all covered under a generic term of delayed failure, slow crack growth, fatigue, creep, or damage initiation/accumulation. Furthermore, additional tests over a wide range of temperatures would be necessary to identify in more detail the failure mechanisms since activation energy could then be established and a temperature-compensated time method could be used to help fit experimental data with an increased accuracy [22].

As shown from the constant stress-rate data presented in this work, high-temperature interlaminar shear strength depends on a test rate that one is to choose when a material exhibits SCG. As a result, the high-temperature strength has a *relative* meaning. Therefore, care should be exercised when the strength is interpreted in terms of its inherent material's parameter. This importance has been addressed for advanced monolithic ceramics [23,24] and CMCs [25,26] in flexure as well as in tension.

Finally, based on the life limiting behavior shown in interlaminar shear, it can be deduced that life limiting may occur in interlaminar tension at elevated temperatures as well since SCG mechanism in tension would be very much reminiscent of that of SCG of monolithic ceramics shown in flexure or in tension [23,24]. An effort to develop high-temperature interlaminar tension test method has been recently made for CMCs and is in progress [27].

Acknowledgements

This work has been sponsored by the Ultra-Efficient Engine Technology (UEET) Program, NASA Glenn Research Center, Cleveland, Ohio. This work is also gratefully supported by the Office of Naval Research and Dr. D. Shifler.

REFERENCES

1. P. Brondsted, F. E. Heredia, and A. G. Evans, "In-Plane Shear Properties of 2-D Ceramic Composites," *J Am Ceram Soc*, **77**[10] 2569-2574 (1994).
2. E. Lara-Curzio and M. K. Ferber, "Shear Strength of Continuous Fiber Ceramic Composites," ASTM STP 1309, p. 31, American Society for Testing & Material, West Conshohocken, PA (1997).
3. N. J. J. Fang and T. W. Chou, "Characterization of Interlaminar Shear Strength of Ceramic Matrix Composites," *J Am Ceram Soc*, **76**[10] 2539-2548 (1993).

4. Ö. Ünal and N. P. Bansal, "In-Plane and Interlaminar Shear Strength of a Unidirectional Hi-Nicalon Fiber-Reinforced Celsian Matrix Composite," *Ceramics International*, **28** 527-540 (2002).
5. S. R. Choi and N. P. Bansal, "Shear Strength as a Function of Test Rate for SiC$_f$/BSAS Ceramic Matrix Composite at Elevated Temperature," *J Am Ceram Soc*, **87**[10] 1912-1918 (2004).
6. S. R. Choi, N. P. Bansal, A. M. Calomino, and M. J. Verrilli, "Shear Strength Behavior of Ceramic Matrix Composites at Elevated Temperatures," *Advances in Ceramic Matrix Composites X*, edited by J. P. Singh, N. P. Bansal, and W. M. Kriven, The American Ceramic Society, Westerville, Ohio; Ceramic Transactions, **65** 131-145 (2005).
7. S. R. Choi, R. W. Kowalik RW, D. J. Alexander, and N. P. Bansal, "Assessments of Life Limiting Behavior in Interlaminar Shear for Hi-Nic SiC/SiC Ceramic Matrix Composite at Elevated Temperature," *Ceram Eng Sci Proc*, **28**[2] 179-189 (2007).
8. S. R. Choi, R. W. Kowalik, D. J. Alexander, and N. P. Narottam, "Elevated-Temperature Stress Rupture in Interlaminar Shear of a Hi-Nic Si/SiC Ceramic Matrix Composite," *Comp Sci Tech*, **69** 890-897 (2009).
9. M. B. Ruggles-Wrenn MB and P. D. Laffey, "Creep Behavior in Interlaminar Shear of Nextel720/Alumina Ceramic Composite at Elevated Temperature in Air and in Steam," *Comp Sci Tech*, **68**[10-11] 2260-2266 (2008).
10. D. Brewer, "HSR/EPM Combustor Materials Development Program," *Mater Sci Eng A*, **261** 284-291 (1999).
11. D. W. Worthem, "Thermomechanical Fatigue Behavior of Three CFCCs," NASA CR-195441, National Aeronautics & Space Administration, Glenn Research Center, Cleveland, OH (1995).
12. S. R. Choi, D. C. Faucett, and D. J. Alexander, "Foreign Object Damage in An N720/Alumina Oxide/Oxide Ceramic Matrix Composite," *Ceram Eng Sci Proc*, **31**[2] 221-232 (2010).
13. ASTM C 1425, "Test Method for Interlaminar Shear Strength of 1-D and 2-D Continuous Fiber-Reinforced Advanced Ceramics at Elevated Temperatures," *Annual Book of ASTM Standards*, Vol.15.01, American Society for Testing & Materials, West Conshohocken, PA (2010).
14. ASTM C 1368, "Standard Test Method for Determination of Slow Crack Growth Parameters of Advanced Ceramics by Constant Stress-Rate Flexural Testing at Ambient Temperature," *Annual Book of ASTM Standards*, Vol. 15.01, American Society for Testing and Materials, West Conshohocken, PA (2010).
15. ASTM C 1465, "Standard Test Method for Determination of Slow Crack Growth Parameters of Advanced Ceramics by Constant Stress-Rate Flexural Testing at Elevated Temperatures," *Annual Book of ASTM Standards*, Vol. 15.01, American Society for Testing and Materials, West Conshohocken, PA (2010).
16. S. R. Choi and N. P. Bansal, "Interlaminar Tension/Shear Properties and Stress Rupture in Shear of Various Continuous Fiber-Reinforced Ceramic Matrix Composites," *Ceramic Transactions,* **175** 119-134 (2006).
17. S. R. Choi and R. W. Kowalik, "Interlaminar Crack Growth Resistance of Various Ceramic Matrix Composites at Ambient Temperature," *J Eng Gas Turbines & Power*, **130**, 031301 (2008).
18. H. Tada, P. C. Paris, and G. R. Irwin, The Stress Analysis of Cracks Handbook, p. 418, ASME, NY (2000).
19. C. A. Lewinsohn, C. H. Henager, and R. H. Jones, "Environmentally Induced Time-Dependent Failure Mechanism in CFCCS at Elevated Temperatures," *Ceram Eng Sic Proc*, **19**[4] 11-18 (1998).

20. C. H. Henager and R. H. Jones, "Subcritical Crack Growth in CVI Silicon Carbide Reinforced with Nicalon Fibers: Experiment and Model," *J Am Ceram Soc*, **77**[9] 2381-2394 (1994).
21. S. M. Spearing, F. W. Zok, and A. G. Evans, "Stress Corrosion Cracking in A Unidirectional Ceramic-Matrix Composite," *J Am Ceram Soc*, **77**[2] 562-570 (1994).
22. J. A. DiCalo, "Creep and Rupture Behavior of Advanced SiC Fibers," Proc. ICCM-10, 6, 315 (1995).
23. S. R. Choi and J. A. Salem, "'Ultra'-Fast Fracture Strength of Advanced Ceramics at Elevated Temperatures," *Mater Sci Eng* A, **242**[1-2] 129-136 (1998).
24. S. R. Choi and J. P. Gyekenyesi, "Elevated-Temperature, 'Ultra'-Fast Fracture Strength of Advanced Ceramics: An Approach to Elevated-Temperature "Inert" Strength," *J Eng Gas Turbines & Power*, **121**[1] 18-24 (1999).
25. *Idem*, "Load-Rate Dependency of Ultimate Tensile Strength in Ceramic Matrix Composites at Elevated Temperatures," *Int J Fatigue*, **27** 503-510 (2005).
26. S. R. Choi, N. P. Bansal, and M. J. Verrilli, "Delayed Failure of Ceramic Matrix Composites in Tension at Elevated Temperatures," *J Euro Ceram Soc*, **25**[9] 1629-1636 (2005).
27. T. Engel, W. Steffier, "High-Temperature Interlaminar Tension Test Method Development for Ceramic Matrix Composites," *Processing & Properties of Advanced Ceramics & Composites III*, The American Ceramic Society, Westerville, Ohio; Ceramic Transactions, **225** (2011).

Nano-Ceramics
and Composites

EFFECT OF COATING PARAMETERS ON THE ELECTRODEPOSITION OF NICKEL
CONTAINING NANO-SIZED ALUMINA PARTICLES

R. K. Saha*, S. Mohamed, and T. I. Khan

University of Calgary
Department of Mechanical and Manufacturing Engineering
2500 University Drive, N.W. Calgary, Alberta T2N 1N4, Canada

ABSTRACT

Composite coatings were developed by co-depositing nano-size (50 nm) Al_2O_3 particles with nickel using the electrodeposition technique. This research investigated the effect of coating parameters e. g. deposition time, pH of the electrolyte, substrate surface roughness and stirring rate on the microstructure, composition, thickness, and mechanical properties of Ni-Al_2O_3 composite coatings. The microstructural features and the mechanical properties of the coated layers were characterized by scanning electron microscopy, Vickers micro-hardness indentation tests, tensile and wear testing. The results showed that the thickness of composite coatings increased with increasin deposition time. Electrolyte with a pH value of 4 and a stirring speed of 600 rpm was most effective in co-depositing the maximum amount of nano-particles in the coatings. The micro-hardness and wear resistance were superior for the composite coatings compared to the pure nickel coating.

1. INTRODUCTION

Nickel coatings have been commercially used as a surface finishing process to modify component surfaces for many years. However, the wider use of pure nickel coatings is limited due to its higher ductility and low hardness. In order to enhance the mechanical properties e. g. hardness, wear and corrosion resistance of the nickel based coatings industries have worked to develop nickel based composite coatings in the recent years. Composite coatings are formed by co-depositing fine inert particles in a metal matrix using various coating techniques among which electrodeposition is a widely used technique. Electrodeposition offers some distinct advantages over other coating methods such as precisely controlled near room temperature operation, low energy requirements, rapid deposition rates, capability to handle complex geometries, low cost, and simple scale-up. Research suggests that the incorporation of various ceramic particles such as Cr_2O_3, SiC, Al_2O_3 and Si_3N_4 in metal matrix can considerably decrease the corrosion rate [1-4]. Furthermore, Cr-Al_2O_3, Ni-SiC, Cu-P, and Ni-P composite coatings are mainly used to increase the wear resistance of metal surfaces [5-10]. Bhagwat et al. [11] and Kuo et al. [4] reported that both wear and corrosion resistant properties increased for a metal surface coated with Ni-Al_2O_3 composite coatings. This was due to the fact that nickel is a corrosion resistant metal and the higher hardness of Al_2O_3 (2720 HV) produces a Ni-Al_2O_3 composite coating that offers improved mechanical properties.

In the fabrication of composite coatings, the amount of strengthening particles and the presence of a uniform dispersion in the metal matrix is a crucial factor which determines the final mechanical properties of the coatings. Earlier research showed that the mechanical properties of composite coatings can be enhanced with an increase in particle concentration in the coatings [12]. However, the production of composite coatings having higher amount of particles with an·even distribution is challenging and the research literature shows that a very low concentration of co-deposited particles in the composite coatings can be achieved [13-19]. Researchers have tried different methods to increase the amount of co-deposited particles in the composite coatings, such as the addition of metal cationic accelerants [20-21] and organic surfactants in an electrolytic bath [22-23], the ultrasonic irradiation [24], and changing the type of applied current [25-26]. Among these methods, use of anion, cation, and nonionic surfactants are used extensively to obtain more incorporation of particles in composite coatings. However, excessive surfactants will reduce the cathode area and increase the brittleness of the coatings [27]. In recent years, research on the formation of Ni-Al$_2$O$_3$ composite coatings [28-32] have focused on the effect of particle size, amount of particles in the electrolyte on the dispersion behavior of ceramic particles in the coatings. However, the quality of the electrodeposited composite coatings will also depend on the coating parameters such as current density, deposition time, pH of the electrolyte, stirring rate, and surface roughness of the substrate used during the electrodeposition process. There is very little research available on the effect of these parameters on the formation of nickel based composite coatings reinforced with nano-sized Al$_2$O$_3$ particles by electrodeposition.

The present work aimed to develop nickel based composite coatings incorporating nano-sized (50 nm) Al$_2$O$_3$ particles. It is thought that the nano-sized particles are lighter in weight, and therefore, more prone to be suspended in the electrolyte during the coating process and this could help achieve a more uniform distribution with a greater concentration of particles in the coatings. The paper focuses on the effects of deposition time, pH of the electrolyte, stirring rate, and surface roughness of the substrate on the morphology of the composite coatings, and the hardness and wear resistance of the Ni-Al$_2$O$_3$ coatings are evaluated. Furthermore, the durability of any coating material will depend not only on the wear resistance of the coatings, but also on the adhesion to the substrate, and therefore, it is important to evaluate the adhesive strength of the coatings [33]. In this study the adhesive strength of the Ni-Al$_2$O$_3$ composite coatings was also evaluated.

2. EXPERIMENTAL PROCEDURE

An electrolysis setup was designed for electrodeposition purpose in which 500 ml of the coating mixture was electrolyzed at room temperature with the DC power supply using nickel and C-Mn based steel (AISI 1018) bars as anode and cathode, respectively. Both electrodes were cleaned by sonification in acidic and basic cleaning mixtures before immersing them in the electrolytic bath. Nickel sulphate hexahydrate (250 g/l), nickel chloride hexahydrate (45 g/l) and boric acid (35 g/l) were dissolved into distilled water to make the electrolytic bath. The nano-sized (50 nm) Al$_2$O$_3$ powder (50 g/l) was added to the electrolytic bath and the dispersions were allowed to stay for 2 hours at room temperature with constant stirring to allow thorough mixing before using the suspension for electrodeposition experiments. The coatings were prepared with a current density of 1 A/dm^2. The selection of current density (1 A/dm^2), particle size (50 nm) and the amount of particles (50 g/l) were based on earlier research work [12, 31, 32]. The effect of other coating conditions e.g. deposition time, pH of the electrolytic bath, stirring rate and the roughness of the substrate surface on the quality of the coatings were also investigated.

Scanning electron microscopy (SEM, JEOL JXA-8200) was used to inspect the surface morphology of the Ni-Al$_2$O$_3$ composite coatings. The coatings were examined using wavelength dispersive spectroscopy (WDS) and energy dispersive spectroscopy (EDS) to determine the weight percent of alumina particles incorporated in the nickel matrix. The aluminum signal obtained by EDS analysis was used to establish the stoichiometric ratio of aluminum to oxygen (2:3) as defined by the chemical formula Al$_2$O$_3$. Vickers hardness of the coated samples was obtained using a micro-hardness tester (Leitz Mini-Load 7840) with an applied load of 50 g. A reciprocating pin-on-disc (POD) tribometer using a silicon nitride ball as the counter-face was employed for the wear tests. All the tests were conducted at room temperature under an applied load of 105g for a total sliding distance of 6 m with a sliding velocity kept at 0.02 m/s. The adhesive strength of electrodeposited coatings was measured following the ASTM C-633 test method.

3. RESULTS AND DISCUSSION

3.1. Effect of deposition time

In order to study the effect of deposition time on the formation of Ni-Al$_2$O$_3$ coatings, the electrodeposition was carried out for 30, 60, 120 and 180 minutes. The other coating parameters were a current density of 1 A/dm2, a temperature of 24°C, a pH of 4, and a stirring rate of 600 rpm. Figure 1 shows the SEM image of the surface of a Ni-Al$_2$O$_3$ coating deposited for 180 minutes. The appearance of the coatings deposited for 30, 60, 120 and 180 minutes were, in general, similar and an even distribution of Al$_2$O$_3$ particles in the nickel matrix was produced for all the coatings. The EDS compositional analysis taken from the cross-section of coatings deposited for 60 minutes detected a 4.3 wt.% of Al$_2$O$_3$ dispersion in the nickel matrix. The amount of Al$_2$O$_3$ particles deposited within the Ni coatings was similar for the various deposition times used in this study. This suggested that the concentration of co-deposited Al$_2$O$_3$ particles in the composite coatings is independent of the deposition time. However, the deposition time had a significant effect on the thickness of the coatings and the coat thickness increased with an increase in deposition time. The coat thickness was measured from the cross-section of the deposited samples using SEM. Figure 2 shows the relationship between the deposition time and the thickness of the Ni-Al$_2$O$_3$ composite coatings.

Figure 1.SEM image of surface appearance of Ni-Al$_2$O$_3$ coatings, deposition time 60 min.

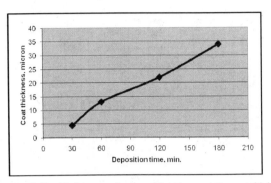

Figure 2. Relationship between the deposition time and the coat thickness.

3.2. Effect of pH

The pH of the electrolyte influences the hydrogen discharge potential at the cathode, the precipitation of basic inclusions, and the composition of the complex or hydrate from which the metal is deposited. Since it is not possible to predict these factors, the best pH range must be determined empirically for specific coating composition. Kim et al. suggested that a pH smaller than 3 should be avoided since nickel sulfamate forms nickel ammonium sulfate which has less solubility below the value of 3 [34]. In this study, coatings were deposited using a pH of 3, 4 and 5 and the effects on the quality of coatings were investigated using constant condition of a current density of 1 A/dm2, a temperature of 24°C, a deposition time of 60 minutes, and a stirring rate of 600 rpm. Figure 3 shows the SEM cross-section image of a Ni-Al$_2$O$_3$ composite coating produced with a pH of 4. The micrograph shows dark dispersions uniformly distributed all over the coatings, and WDS analysis taken from these dispersions revealed the dark areas to be rich in aluminum, see figure 4. However, the concentration of Al$_2$O$_3$ particles in the coatings varied with a change in the pH of the electrolytic bath. As can be seen in figure 5, the amount of Al$_2$O$_3$ particles in the composite coatings was highest at pH 4. A similar result was reported by Kim et al. for the formation of Ni-SiC composite coatings by electrodeposition technique [34]. When fabricating nickel based composite coatings with a dispersion of diamond particles, Moon et al. suggested that at low pH the diamond particles were not co-deposited and the contents of diamond increased with the increase of pH of the electrolytic bath [35]. The effect of pH on the deposition rate for nickel-alumina coatings was studied by Marikkannu et al. and a pH value of 5-6 was found more productive for their coating conditions [36]. The pH of electrolytic bath plays an important role in changing the efficiency of the electrodes e. g. cathode and anode. Cathode efficiency is defined as the ratio of the actual amount of the deposited material to the theoretical amount that should be deposited, whereas anode efficiency is the ratio of the actual amount of the dissolved material to the theoretical amount that should be dissolved. Increasing pH means that cathode efficiency is lower than anode efficiency; decreasing pH, the reverses this effect. In the current study, a balance in electrode efficiencies was found to be achieved at a pH of 4 for the electrodeposition conditions described in experimental procedure. This observation points to the fact that in the given composition and coating conditions, a pH of 4 was more effective in mobilizing the positively charged Al$_2$O$_3$ particles along with the nickel cations towards the cathodic substrate. A coat thickness of 20, 22.5 and 21 μm was measured for the composite coatings made at a pH of 3, 4 and 5, respectively.

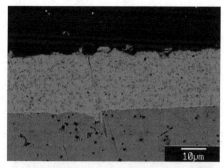

Figure 3.SEM image of cross-section of Ni-Al$_2$O$_3$ composite coatings deposited at a pH of 4.

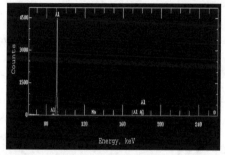

Figure 4.WDS compositional analysis taken from the dark dispersion shown in figure 3.

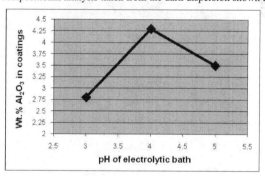

Figure 5.Effect of the pH on the concentration of Al$_2$O$_3$ particles in the coatings

3.3. Effect of stirring speed

In the formation of composite coatings by electrodeposition, inert particles must be transported to the cathode substrate for the co-deposition process, and the stirring speed affects the concentration of particle in the coatings. The stirring speed becomes even more important when light-weight, nano-sized particles are used to form composite coatings. In order to study the effect of stirring speed on the dispersion behavior of the Al_2O_3 particles, coatings were produced with a stirring speed of 200, 400, 600 and 800 rpm. The coating conditions were a current density of 1 A/dm2, a temperature of 24°C, a deposition time of 60 minutes, and a pH of 4. Figure 6 shows the relationship between the stirring speed and the wt% of Al_2O_3 particles in the composite coatings. The graph shows that the amount of co-deposited particles increased with an increase in stirring speed and the highest concentration of Al_2O_3 particles was recorded at 600 rpm beyond which the amount of particles deposited decreased. This finding was consistent with earlier observation reported by Kim et al. for the formation of Ni-SiC composite coatings by electrodeposition technique [34]. Badarulzaman et al. found that the concentration of embedded alumina particles in the nickel matrix increased with an increase in stirring speed [37]. However, a much lower range of stirring speed (100–250 rpm) was employed in their study where the concentration of Al_2O_3 particles in the electrolytic bath was only 2 g/l. In the current study the concentration of Al_2O_3 particles in the electrolytic bath was 50 g/l and a higher stirring speed were necessary to keep the particles suspended. When the coatings were performed at a slow stirring speed of 200 rpm, a large amount of particles were found to settled at the bottom of the electrolyte. On the other hand, the higher stirring speed tends to wash away some of the adsorbed Al_2O_3 particles from the cathode substrate due to the particle-coating friction which results in a decrease in the amount of co-deposited particles.

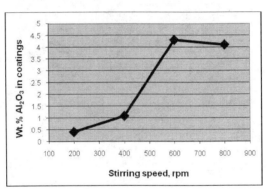

Figure 6.Effect of stirring rate on the concentration of Al_2O_3 particles in the coatings

3.4. Effect of substrate surface roughness

In order to study the effect of surface roughness, substrates were prepared with a roughness of 0.42, 0.26, 0.18 and 0.1 μm, and coatings were deposited onto these substrates using a current density of 1 A/dm2, a temperature of 24°C, a deposition time of 60 minutes, a pH of 4, and a stirring rate of 600 rpm. Micrographs in Figure 7 show the surface morphology of the coatings deposited on the substrate with a surface roughness of 0.42 and 0.1 μm. As can be seen from these micrographs, the coatings were smooth with a minimal defect when deposited on a fine substrate surface. Coatings deposited on a rough surface experienced some micro-cracks which could be attributed to the internal stresses associated with the evolution of hydrogen bubbles. It is thought that the stirring motion of the electrolyte failed to remove bubbles entrapped in the deeper asperities of the rough cathode surface, while gas bubbles are easily removed from the smooth surface of the substrate.

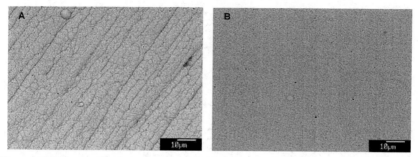

Figure 7. SEM image of surface appearance of Ni-Al$_2$O$_3$ coatings deposited on the substrate with a surface roughness of (A) 0.42 μm and (B) 0.1 μm

3.5. Mechanical properties of the Ni-Al$_2$O$_3$ coatings

It is well know that the mechanical properties of the composite coatings largely depend on the concentration of strengthening particles in the metal matrix. The presence of a fine uniform dispersion of particles obstructs the movement of dislocations and resists plastic flow. Therefore, an increase in particle concentration within the coatings can enhances the mechanical properties of the coatings. In this study, the best co-deposition in terms of particle concentration (4.3 wt% Al$_2$O$_3$) was achieved with the following coating parameters: deposition time – 60 minutes, pH – 4, stirring speed – 600 rpm, and this sample was chosen for the measurement of mechanical properties e. g. micro-hardness, wear resistance and the bond strength.

Micro-hardness: To assess the uniformity of particle dispersion in the coatings, micro-hardness was measured through cross-sections of the coatings. Five measurements were recorded from different area of the coatings and the values were found to be close to each other which suggested that the particle distribution was uniform. The average hardness value for a particle concentration of 4.3 wt% in the coatings was measured to be 726 HV.

Bond strength: Adhesive strength of the electrodeposited coating was evaluated from the load at which fracture occurred between the coated specimen and the adhesively bonded reference sample. The adhesive strength of the resin, provided by the manufacturer, was 80 MPa. The average bond strength of the coatings produced with a particle concentration of 4.3 wt% was measured to be 62 MPa. Figure 8 shows the image of the coated and reference specimens after the tensile test. As can be seen from this figure, a partial removal of the coating from the steel substrate to the reference specimen took place. The portion of the coating present with the substrate specimen is shown with the arrow marks in figure 8. This result suggested that the fracture can fall in a dual-breaking region which means the specimen might break at either coating/substrate interface or within the resin.

Figure 8. Optical microscopic image of the Ni-Al₂O₃ coated sample after the tensile test: (A) reference specimen; (B) coated specimen.

Wear resistance: The coefficient of friction for a silicon nitride ball sliding against the Ni-Al₂O₃ composite coating was measured and the results were compared with that of a pure nickel coating. Figure 9 shows the variation in friction coefficient for a coating containing 4.3 wt% Al₂O₃ and a pure nickel coating. As can be seen from this figure, the coefficient of friction was initially low at the beginning of the test, but began to increase as the wear test progressed. This change with increasing sliding distance was attributed to a change in the state of the coating surface due to the increasing plastic deformation and eventual delamination of the coating. A large variation in coefficient of friction was evident for both of the samples which indicated a stick-slip behavior for these coatings. The average coefficient of friction of 0.91 and 0.75 was measured for the composite coatings and the pure nickel, respectively.

Figure 10 shows the morphology of the worn surface of the Ni-Al₂O₃ composite coatings with a particle concentration of 4.3 wt% and the pure nickel coating. Both samples suffered plastic deformation and the degree of plastic deformation was higher for the pure nickel coating. The width of the wear track generated on the coated surfaces was taken as a quantitative measure of the wear resistance of the electrodeposited coatings, see figure 10. It can be seen from this micrograph that the width of the wear track was smaller than that obtained on the pure nickel coatings. This suggested that the composite coatings showed better sliding wear resistance than the dispersion free coatings.

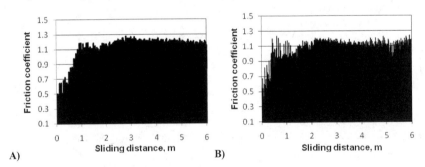

Figure 9.Coefficient of friction for electrodeposited coatings: (A) Ni-Al₂O₃ composite coating with a particle concentration of 4.3 wt.%, (B) pure nickel coating

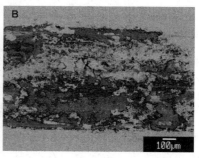

Figure 10.SEM image of the worn surface of the coatings: (A) Ni-Al$_2$O$_3$ composite coating with a particle concentration of 4.3 wt%, (B) pure nickel coating

4. CONCLUSIONS

Nickel coatings containing a dispersion of nano-sized Al$_2$O$_3$ were successfully produced by electrodeposition. The effect of coating parameters (deposition time, pH of electrolyte, stirring rate and substrate surface roughness) on the quality and mechanical properties of the coatings were studied. The deposition time had a significant impact on the thickness of the deposited film and the coat thickness increased with an increase in the deposition time. However, the maximum amount of Al$_2$O$_3$ particles (4.3 wt%) was achieved when the coating was deposited for 60 minutes with a pH of 4 and a stirring speed of 600 rpm. The hardness and the bond strength for this sample were measured to be 726 HV and 62 MPa, respectively. The incorporation of nano-sized particles increased the sliding wear resistance of the coatings.

ACKNOWLEDGEMENTS

The authors would like to acknowledge Alberta Ingenuity Fund, Canada for the financial support for this work.

FOOTNOTES

* Corresponding author (E-mail: rksaha@ucalgary.ca)

REFERENCES

1. J. P. Celis, and J. R. Ross, *J. Electrochemical Society*, **124**(10), 1508-511 (1977).
2. G. N. K. Ramesh Bapu, and M. Mohammed Yusuf, *Mater Chem Phys*, **36**(1), 134-38 (1993).
3. Y. Y.Cheng, *Masters Thesis,* Chung-Cheng Institute of Technol., Taoyuan, Taiwan, (1996).
4. S. L. Kuo, Y. C. Chen, W. H. Hwu, M. D. Ger, and X. L. Chen, *J. Chinese Institute of Chemical Engineers*, **34**(4), 393-98 (2003).
5. R. Narayan, and S. Chattopadhyay, **16**(3), 227-34 (1982).
6. J. W. Graydon, and D. W. Kirk, *J. Electrochemical Society*, **137**(7), 2061-066 (1990).
7. N. Periene, A. Cesuniene, and L. Taicas, *Plating Surface Finishing*, **81**(10), 68-71, (1993).

8. G. Maurin, and A. Lavanant, *J. Apply Electrochemistry*, **25**(12), 1113-121 (1995).
9. J. R. Roos, J. P. Celis, J. Fransaer, and C. Buelens, *J. Metals*, **42**(11), 60-63 (1990).
10. K. N. Sun, X. N. Hu, J. H. Zhang, and J. R. Wang, *Wear*, **196**(1), 295-97 (1996).
11. M. J. Bhagwat, J. P. Celis, J. R. Roos, *Trans. Institute of Metal Finish.*, **61**(2), 72-79 (1983).
12. R. K. Saha, and T. I. Khan, *Surf. Coat. Technol.*, DOI:10.1016/j.surfcoat.2010.08.035 (2010).
13. J. Foster, and B. Cameron, *Trans. Institute of Metal Finish.*, **54**(4), 178-83 (1976).
14. L. Benea, P. L. Bonora, A. Borello, S. Martelli, F. Wenger, P. Ponthiaux, and J. Galland, *J. Electrochemical Society*, **148**(7), 461-65 (2001).
15. A. B. Viderine, E. J. Podlaha, *J. Applied Electrochemistry*, **31**(4), 461-68 (2001).
16. I. Garcia, J. Fransaer, and J. P. Celis, , *Surf. Coat. Technol.*, **148**(2-3), 171-78 (2001).
17. J. Fransaer, J. P. Celis, and J. R. Ross, *J. Electrochemical Society*, **139**(2), 413-25 (1992).
18. I. Shao, P. M. Vereecken, R. R. Cammarata, and P. C. Searson, *J. Electrochemical Society*, **149**(11), 610-14 (2002).
19. S. C. Wang, and W. C. J. Wei, *Mater. Chem. Phys.*, **78**(3), 574-80 (2003).
20. L. Wang, Y. Gao, H. Liu, Q. Xue, and T. Xu, *Surf. Coat. Technol.*, **191**, 1-6 (2005).
21. C. S. Lin, and K.C. Huang, *J. Appl. Electrochem.*, **34**, 1013-019 (2004).
22. L. Chen, L.Wang, Z. Zeng, and J. Zhang, *Mater. Sci. Eng. A*, **434**, 319-25 (2006).
23. M.D. Ger, *Mater. Chem. Phys.*, **87**, 67-74 (2004).
24. D. Lee, Y.X. Gan, X. Chen, and J.W. Kysar, *Mater. Sci. Eng. A*, **447**, 209-16 (2007).
25. L. Chen, L. Wang, Z. Zeng, and T. Xu, *Surf. Coat. Technol.*, **201**, 599-605 (2006).
26. L.M. Chang, M.Z. An, H.F. Guo, and S.Y. Shi, *Appl. Surf. Sci.*, **253**, 2132-137 (2006).
27. A. Hovestad, R.J.H.L. Heesen, and L.J.J. Janssen, *J. Appl. Electrochem.*, **29**, 331-38 (1999).
28. L. Du, B. Xu, S. Dong, H. Yang, and Y. Wu, *Surf. Coat. Technol.*, **192**, 311-16 (2005).
29. S. Kuo, Y. Chen, M. Ger, and W. Hwu, *Mater. Chem. Phys.*, **86**, 5-10 (2004).
30. S. Wang, and W. J. Wei, *Mater. Research*, **18**, 1566-574 (2003).
31. R. K. Saha, T. I. Khan, L. B. Glenesk, and I. U. Haq, *Ceramics Trans.*, **208**, 37-44 (2009).
32. R. K. Saha, I. U. Haq, T. I. Khan, and L. B. Glenesk, *Key Eng. Mater.*, **442**, 187-94 (2010).
33. T. Arai, H. Fujita, and M. Watanabe, *Thin Solid Films.*, **154** 387-401 (1987).
34. S. K. Kim and H. J. Yoo, *Surf. Coat. Technol.*, **108-109**, 564-69 (1998).
35. Y. Moon, J. Lee, T Oh, and J. Byun, *Key Eng. Mater.*, **345-346**, 1597-1600 (2007).
36. K. R. Marikkannu, K. Mutha, G. P. Kalaignan, and T. Vasudevan, in Proc. Int. Symp. of Research Students on Mater. Sci. and Eng., Chennai, India, 20-22 December 2004.
37. N. A. Badarulzaman, S. Purwadaria, A. A. Mohamad, and Z. A. Ahmad, *Ionics*, **15**, 603-07 (2009)

SYNTHESIS OF Al$_2$O$_3$-TiC NANO-COMPOSITE PARTICLES BY A NOVEL ELECTRO-PLASMA PROCESS

Kaiyang Wang [a), b)], Peigen Zhang [a)], Jiandong Liang [a)] and S.M. Guo [a)]
[a)] Department of Mechanical Engineering, Louisiana State University, Baton Rouge, LA 70803

[b)] CFAW Ceramics LLC, Baton Rouge, LA 70820

ABSTRACT

The formation of nano-sized alumina–titanium carbide (Al$_2$O$_3$-TiC) composite powders from aluminum, graphite and nano-sized anatase titania powder compact were directly synthesized by electro-plasma process (EPP) and the following combustion reaction. The theoretical density of compact played an important role for the ignition of the combustion reaction. For the compact with higher than 80% of theoretical density, the combustion reaction occurred and the nano-composites of Al$_2$O$_3$-TiC were formed in-situ. Because of the occurrence of combustion reaction in the electrolyte liquid, the resultant particles experienced rapid quenching and resulted in ultrafine microstructures (50 – 100nm). X-ray diffraction (XRD) and scanning electron microscopy (SEM) were used to characterize the products after the processing by electro-plasma. The mechanism of formation of nano-sized alumina and titanium carbide was also discussed.

INTRODUCTION

Ceramic – ceramic composites of TiC–Al$_2$O$_3$ have been the focus of significant attention primarily because of their potential applications in cutting tools and tape-head materials [1, 2, 3]. The utilization of TiC–Al$_2$O$_3$ in these and similar applications is motivated by the attractive properties such as high wear resistance, high strength and fracture toughness, and good electrical conductivity. At present time, commercial TiC–Al$_2$O$_3$ ceramic composites are primarily manufactured by pressureless sintering or hot-pressure of TiC and Al$_2$O$_3$ powders whose composition is usually 30 wt% of TiC in the mixture.

Self-propagating high-temperature synthesis (SHS) or combustion synthesis (CS) was first developed by the Institute of Structural Macrokinetics and Materials Science (ISMAN) in Russia in the late 1960's [4]. This technique employs exothermic reaction processing, which circumvents the difficulties of long processing time and high energy consumptions associated with the conventional sintering method. The combustion synthesis of Al$_2$O$_3$–TiC composites can be achieved by the following reaction:

$$3TiO_2 + 4Al + 3C = 2Al_2O_3 + 3TiC \qquad (1)$$

which is strongly exothermic, having an adiabatic temperature of 2329 K. Under these conditions, one of the product phases, Al$_2$O$_3$, is molten (melting point: 2288 K) while the other, TiC, is solid (melting point: 3313 K). The combustion reaction of Eq. (1) self-propagates at a relatively high rate despite its complex nature [5]. For this process, the generally accepted mechanism has two sequential reactions: a thermite reaction between the aluminum and the titanium oxide followed by a reaction between the graphite and the titanium metal, i.e.,

$$3TiO_2 \lessgtr 4Al \ \nabla \ 2Al_2O_3 \lessgtr 3Ti \qquad (2)$$
$$Ti \lessgtr C \ \ \nabla \ \ TiC \qquad (3)$$

In general, there are two modes by which combustion synthesis may occur: self-propagating high-temperature synthesis (SHS) and volume combustion synthesis (VCS) [6]. In both cases, reactants may be pressed into a pellet, typically cylindrical or parallelepiped – shaped. The samples are then heated by an external source (e.g. electric arc [5], tungsten coil [7], etc.), either locally (SHS) or uniformly (VCS), to initiate an exothermic reaction. Using SHS and VCS, the typical Al_2O_3 and TiC particles sizes are 5 ~ 20 μm and 2 ~ 10 μm, respectively [7].

Electrochemical discharges also known as contact glow discharge electrolysis, is a well known phenomenon and has been systematically investigated [8, 9, 10] in the past. Oishi et al. [11] showed the fabrication of metal oxide nano-particles by anode discharges between a metal anode and molten salt electrolyte in argon atmosphere at 700 K by applying 500 V DC. Toriyabe et al. [12] reported the controlled formation of metallic nanoballs during plasma electrolysis. Liang et al. [13] reported to use EPP for surface cleaning. A gas/vapor sheath is formed at the solid electrode/electrolyte interface when the applied voltage is high enough to induce discharge plasma. In this paper, in order to synthesize Al_2O_3–TiC composite nano-particles, a novel process, which involves electro-plasma, combustion reaction, and liquid quenching, was developed. The metal-ceramics composites powder compact was used as a cathode. Through controlling electrolyte concentration, applied DC voltage and the compacting pressure, electro-plasma discharge occurred around the aluminum-graphite-titania composites surface and induced the combustion reaction to produce alumina–titanium carbide nano-particles.

EXPERIMENTAL PROCEDURES

The reactants were powders of titania (99% pure, anatase, average particle size of 0.5 μm), aluminum (99% pure, atomized, average particle size of 30 μm), and graphite (99% pure, average particle size of 60 μm). Appropriate amounts of these powders were mixed in accordance with the stoichiometric ratios of Eq. (1). Mixing was done in a stainless steel ball-mill using stainless steel balls for 20 minutes in an argon atmosphere under cryogenic temperature and the blends were then pressed into compacted electrode samples. The dimension of compact electrode was 12.5 mm in diameter and 2.5 mm in thickness, respectively. To yield 65, 75, 85 % of the theoretical density, the press machine was set to be 80, 120, and 160 MPa respectively.

The electro-plasma apparatus was set-up as shown in Figure 1. The water based electrolyte solution was prepared with 75g/L sodium carbonate, which gave an equivalent electrical conductivity of 43±0.5 mS/cm. During the experiment, there is no preheating to the electrolyte bath. A DC power source (Magna-power PQA) was connected to the powder compact cathode and an AISI 304 stainless steel plate anode to form a complete circuit. The input voltage was set to be 200 V, which took 2±0.5 seconds to reach from zero. The cathode was fed by an electric motor at a constant speed (1.0 mm/min) toward the electrolyte bath. As the cathode just made contact with the electrolyte, electro-plasma formed between the electrolyte and the contacting cathode surface. Due to electrical heating, electrolyte turns into liquid-vapor two phases and instability occurs. High potential between the electrodes leads to concentration of positive ions that are present in the electrolyte, in the close proximity of the cathode, mostly on the surface of the gas bubbles. This results in high localized electric field strength between the cathode and the positive charges. When the electric field strength reaches ~10^5 V/m or higher, gas space inside the bubbles is ionized and a plasma discharge takes place. In this process, the cathode was continually fed toward the electrolyte bath; cathode disintegrated at the contact by electro plasma; and reaction happened in the disintegrated cathode particles in the electrolyte bath. After the reaction, the solution was then diluted and purged with distilled water to remove sodium and carbonate ions. Resultant residual was collected and dried in air for microstructure characterization.

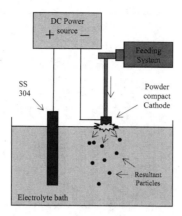

Figure 1 Electro-plasma processing (EPP) set-up

The phases in the product were characterized using a powder X-ray diffraction (XRD) (CuKα, λ=0.154nm, Rigagu). Field emission scanning electron microscopy (FESEM) (Model Quanta 3D FEG, FEI Company, USA) was used to characterize the powder microstructures.

To compare with the EPP derived alumina–titanium carbide composites, a compact pellet was directly ignited by an electric arc. The as-combusted and fractured sample was also examined by XRD and FESEM methods.

RESULTS AND DISCUSSION

For the composite powder compacts, made under different compacting pressures thus with different apparent densities, electro-plasma processing was applied. No reaction was found for the compact samples with a density less than the 80% of theoretic value. Instead, for the samples with lower densities, the powder compacts were disintegrated under the electro-plasma processing. In contrast, for samples with an apparent density over 80% of theoretic density, electro-plasma was successfully maintained around the tip of the compact cathode. At a proper feeding rate, the plasma envelope makes ablation on the cathode and induces the reaction among the falling particles, which are quenched immediately by the surrounding electrolyte liquid. By controlling the electrical current, the core of the cathode may be prevented to a high temperature to induce the combustion reaction.

Figures 2 (a) and (b) show the X-ray diffraction (XRD) patterns of the initial as-milled powder mixtures and the products after electro-plasma processing, with induced reactions and quenching. From Figure 2(a), diffraction peaks of aluminum, anatase TiO_2 and graphite are clearly seen. After electro-plasma processing, titanium carbide and alumina are formed in a complete reaction, Figure 2(b).

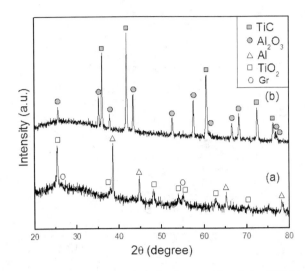

Figure 2 patterns of as-milled powder mixtures (a) and after EPP (b)

FESEM images of the EPP products are shown in Figure 3, which shows the crystal sizes of the particles are below sub-micron range. Micro-elemental (EMPA) analyses on these microstructures show that TiC is partially dissolved in corundum Al$_2$O$_3$ after the reaction. Due to the limitation of EDS resolution, the interfaces between TiC and Al$_2$O$_3$ cannot be identified. However, EDS shows that the large particles contain more Al$_2$O$_3$ and the small particles contain more TiC. The average Al$_2$O$_3$ rich particle size is around 200nm and the average TiC rich particle size is around 50nm. This result is consistent with those reported by Xia et al., which suggested that the Al$_2$O$_3$ was in molten state and TiC was in solid state during this reaction stage.

Figure 4 show the X-ray diffraction (XRD) pattern of the compacted sample (same pressure as above sample) after ignition by electric arc. Titanium carbide and alumina are formed in a complete reaction. We cannot find any residues in this pattern. Figure 5 (a) and (b) shows the FESEM image of the cross-section of the sample after electric arc ignition. One can clearly see that the large grain size powders are Al$_2$O$_3$ and small grain size powders are TiC. Compare to the above resultant powders obtained from electro-plasma processing, the grain sizes of Al$_2$O$_3$ and TiC are obvious larger than those of shown in Figure 3 (a) and (b). The sintering phenomenon of alumina is also very obvious.

Figure 3(a) SEM micrograph of products obtained after electro-plasma processing

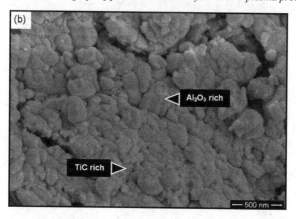

Figure 3(b) High magnification image of Figure 3(a)

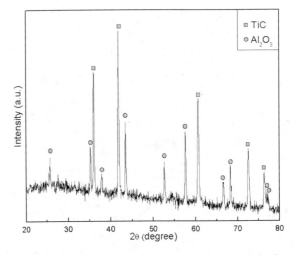

Figure 4 XRD patterns of as-milled powder mixtures compacts after SHS by electric arc ignition

Figure 5 (a) FESEM micrograph of products obtained after SHS by electric arc ignition.

Figure 5(b) FESEM micrograph of products obtained after SHS by electric arc ignition.

Gidalevich et al. [14] reported a theoretical analysis of a radial plasma column induced hydrodynamic effects under pulsed low current arc discharges. It was found that during the initial 10^{-8} s the velocity of the plasma column expansion exceeds many times of the speed of sound, creating a shock wave in the water with a pressure discontinuity of ~10 GPa, and a high temperature in the order of 10^4 to 10^5 K. Timoshkin et al. [15] reported that a high voltage spark discharge could induce a pressure up to several GPa in the plasma bubble. High temperature in plasma bubbles leads to the melting of localized surfaces on the cathode and the high pressure atomizes the melt to small droplets. These droplets are quenched by the surrounding electrolyte, leading to the formation of nano-particles [12].

The present experiment is different from those previously reported ignition methods such as electric arc and tungsten wire [5, 7]. Figure 6 shows the schematic reaction mechanism of forming alumina-titanium carbide composite nano-particles during the EPP treatment. In this process, the bonding strength of the powder compact plays an important role for the completion of following combustion reaction. For the samples with low theoretical densities (i.e. 65% and 75%), the as-milled particles are easily falling off the compact, thus no induced combustion reaction. For samples with high apparent density (80%), the combustion reaction can occur and generate Al$_2$O$_3$-TiC nano-composite powders. Once a highly densified powder compact (cathode) makes contact with electrolyte, large current is introduced through the contact surface due to good electrical conductivity. Consequently, local joule heat evaporates the electrolyte solution. As such, an unstable gas layer, mainly water vapor, is formed around the cathode/electrolyte interface. High voltage application through such gas layer induces localized plasma, which could be as high as 3000-6000 K [16, 17, 18].

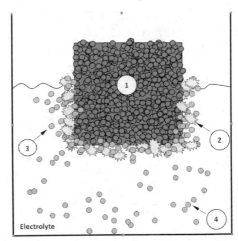

Figure 6 Reaction mechanism of formation of Al2O3 + TiC composite nano-particles by EPP;
(1) powder compact (Cathode); (2)plasma; (3)melted particles; (4)resultant particles

Unlike the traditional electro-plasma process where metallic solid electrodes are used, a combustion synthesis of alumina–titanium carbide nano-particles is obtained using aluminum-graphite-titania powder compacting electrodes in an electro-plasma process. In this process, the compacting electrode is locally partial melted and ejected by the localized micro electro-plasma arcs, induced by electrothermal instability. Combustion reactions happen quickly in the melted composite electrode droplets. With cold surrounding solution, the ejected molten cathode droplets, due to localized shock wave from plasma discharge, experience rapid quenching and result in very fine grain sizes. Comparing to the traditional process of combustion reaction, this EPP-assisted combustion reaction provides fast cooling from ambient solution, by which combustion reaction could be controlled.

CONCLUSION

Al₂O₃ + TiC nanocomposites can be directly synthesized through electro-plasma processing and following combustion reaction and quenching using aluminum, graphite and titania powder compacts. The apparent density of the powder compacts played an important role for the onset of electro-plasma induced combustion/quenching reactions. For the compacts with theoretical density larger than 80%, the combustion/quenching reactions can occur. Because of the occurrence of combustion reaction in the EPP ejected small particles in a liquid environment, the resultant particles experienced rapid quenching and resulted in ultrafine Al₂O₃-TiC nano-composites. Comparing to other methods, such as applying arc discharge in a vacuum chamber, the electro-plasma discharge method presented in this paper can be used to produce nano-composite powders easily, quickly, and inexpensively.

REFERENCES

[1] A. G. King, *Am. Ceram. Soc. Bull.*, 43 (1965)395.

[2] S. Adachi, T. Wada, T. Mihara, Y. Miyamoto, M. Koizumi, *J. Am. Ceram. Soc.* 73(1990) 1451.

[3] Y. Choi, S. W. Rhee, *J. Am. Ceram. Soc.* 78(1995) 986.

[4] V. I. Nikitin, A. I. Chmelevskich. A. P. Amosov, A. G Merzhanov, *Book of Abstracts of the 1st International Symposium on SHS*, Alma-Ata. Russian Federation (1991).

[5] T.D. Xia, Z. A. Munir, Y.L. Tang, W.J. Zhao, T.M. Wang, *J. Am. Ceram. Soc.* 83(2000)507.

[6] A. Varma, A.S. Mukasyan, Korean *J. Chem. Eng.* 21(2004)527.

[7] J.H.Lee, S.K.Ko, C.W.Won, *Mater. Res. Bull.* 36(2001)989.

[8] C. Guilpin, J. Garbaz-Olivier, *Spectrochim. Acta B* 32 (1977) 155.

[9] H. Vogt, *Electrochim. Acta* 42 (1997) 2695.

[10] R. Wüthrich, C. Comninellis, H. Bleuler, *Electrochim. Acta* 50 (2005) 5242.

[11] T. Oishi, T. Goto, Y. Ito, *J. Electrochem. Soc.* 149 (2002) D155.

[12] Y. Toriyabe, S. Watanabe, S. Yatsu, T. Shibayama, T. Mizuno, *Appl. Phys. Lett.* 91(2007) 041501 .

[13] J.D. Liang, M.A. Wahab, S.M. Guo, 2010, Surface Cleaning and Surface Modifications through the Development of a Novel Technology of Electrolytic Plasma Process (EPP), *the World Journal of Engineering*, 2010 (7.3): 54-61.

[14] E. Gidalevich, R. L. Boxman, S. Goldsmith, *J. Phys. D: Appl. Phys.* 37(2004)1509.

[15] I.V. Timoshkin, R.A. Fouracre, M.J. Given, S.J. MacGregor, *J. Phys. D: Appl. Phys.* 39 (2006) 4808.

[16] A.L. Yerokhin, X Nie, A. Leyland, A. Matthews, S.J. Dowey, *Surf. & Coat. Tech.* 122 (1999)73.

[17] P. Gupta, E.O. Tenhundfeld, E.O. Daigle, D. Ryabkov, *Surf. & Coat. Tech.* 201(2007)8746.

[18] M.D. Klapkiv, H.M. Nykyforchyn, V.M. Posuvailo, *Materials Science* 3(1994)333

Testing, Characterization, and Microstructure-Property Relationships

RESEARCH ON TRIBOLOGICAL PROPERTIES OF REACTIVE SINTERED Si_3N_4-BASED COMPOSITE CERAMIC

MA Hai-long[1,2], CUI Chong[1], LI, Xing[3], WANG Yuan-ting[1], ZHOU Hai-jun[1]

(1. School of Materials Science and Engineering, Nanjing University of Science and Technology, Nanjing 210094, China;
2. Institute of Ethnic Preparatory Education, Ningxia University, Yichuan 750021, China;
3. School of Mathematics and Computer Science, Ningxia University, Yichuan 750021, China)

ABSTRACT

Reaction sintered Si_3N_4-based composite ceramic was prepared by quasi-static temperature-rising craft with the powder of Si as raw material, SiC and Al as additives. Results of friction experiment show that the tribological properties of ceramics were influenced by load, lubrication and counterface materials. The steady friction coefficient and wear rate increase with the increase of load. The wear rate of materials decreased under the water-lubricatation condition and the wear property of ceramics was improved. Both the friction coefficient and wear rate of ceramic samples sliding against GCr15 ring were lower than samples sliding against Si_3N_4 ring under the same friction conditions.

INTRODUCTION

In recent years, reactive sintered Si_3N_4-SiC ceramic materials have shown excellent resistance to oxidation, wear and high temperature, but the strength of these materials have remained considerably low. Factors contributing to the weak mechanical properties of Si_3N_4-SiC ceramics include the interface between Si_3N_4 and SiC as well as low volume density. The interface between Si_3N_4 and SiC limits the addition of SiC and increases intergranular fracture tendency of these materials [1]. Reactive sintered Al bonded Si_3N_4-SiC ceramics improve the mechanical properties and the interface combinative modality of the Si_3N_4-SiC ceramic materials. So it is hoped to replace currently used reactive sintered Si_3N_4-based ceramics.

In our previous study, aluminium nitride-silicon nitride-silicon carbide (AlN-Si_3N_4-SiC) composite ceramics were synthesized by reactive sintering in N_2 atmosphere at 1360 °C using Al, Si and SiC as raw materials [2], SEM of aluminium nitride-silicon nitride-silicon carbide composite ceramics (Al:Si:SiC=2:7:3) were shown in Figure 1. The influence of Al addition on the phase composition and microstructure along was also discussed [3]. It was concluded that the interface between Si_3N_4 and SiC gets improved effectively by adding Al to the Si_3N_4-SiC ceramic. The addition of Al was found to mitigate interfacial limitations allowing for the increased SiC into the microstructure. Even though the addition of Al increased SiC content another issue arose. It was also pointed out that liquid Si content increases with Al content at the same temperature and results in formation of brittle phase in the center of ceramic which adversely affects the mechanical properties of these materials. Therefore, the Al content should be controlled. Bending strength degradation occurs when the Al content is higher than 29%. When Al content achieves 38%, the effect of strengthening with addition of Al is not obvious because Si inhibits advancing the strength of materials.

(a) ×2000 (b) ×4000

Figure 1 SEM of aluminium nitride-silicon nitride-silicon carbide composite ceramics

So far, mechanical properties of reactive sintered Si$_3$N$_4$-based composites with the addition of Al have been studied, but not their tribological properties. Therefore, the aim of the present paper is to explore the effect of the addition of Al on the tribological properties of reactive sintered Si$_3$N$_4$-based composites under abrasive conditions. The effect of load, lubrication and materials of friction rings on friction and wear properties has been investigated and discussed.

EXPERIMENTAL

In previous study, Al, SiC and Si powders were used as raw materials; eight different systems of sintering additives were used in that study. Kept the proportion of Si to SiC of 7 to 3, and Al from 0 to 8. The starting powders were mixed for 2h using absolute alcohol solvent in a steel jar with 1Cr18Ni9Ti balls as the milling media. After ball-milling, the slurry was dried by using an evaporator. PVA was used as mixed binder. Dry-pressing experiments were carried out under applied pressure of 100 MPa. Lower temperature nitrification sintering was carried out at 700-1000°C and then reactive sintering were done at 1360°C for 5h. The results showed that samples doped with Al exhibit higher strength at room temperature, the sample (Al:Si:SiC=4:7:3) especially showed average strength of 146 MPa and maximum strength of 179 MPa [1]. Therefore, in order to prepare reactive sintered Si$_3$N$_4$-based ceramics composite materials with better mechanical properties and tribological properties, SiC proportion and then the Aluminum proportions were fixed, the effect of Al proportion and SiC proportion on the performance of reactive sintered Si$_3$N$_4$-based ceramics composite materials will be investigated, respectively, the detailed mix proportion of reactive sintered Si$_3$N$_4$-based ceramics composite materials are given in Table 1 and Table 2.

Specimens of 8mm×10mm×55mm in dimension were prepared for tribological tests (MM-200 friction testing machine, China). Schematic diagram of MM-200 experimental facility is shown in Fig.2. Two different loads, namely 49 and 147 N, were used for the wear testing. Worn surfaces of specimen were observed by scanning electron microscopy (JEOL JSM-63 80LV).

Table 1 Mix proportion of reactive sintered Si_3N_4-based ceramics composite materials

Group No.	Proportion (Al:Si:SiC)			Al (wt %)	Ball-milling time (min)
	Al	Si	SiC		
A1	0	7	0	0	120
A2	0	7	3	0	120
A3	1	7	3	9%	120
A4	2	7	3	17%	120
A5	3	7	3	23%	120
A6	4	7	3	29%	120
A7	5	7	3	33%	120
A8	6	7	3	38%	120
A9	7	7	3	41%	120
A10	8	7	3	44%	120

Table 2 Mix proportion of reactive sintered Si_3N_4-based ceramics composite materials

Group No.	Proportion (Al:Si:SiC)			SiC (wt %)	Ball-milling time (min)
	Al	Si	SiC		
B1	4	7	0.5	4%	120
B2	4	7	1	8%	120
B3	4	7	1.5	12%	120
B4	4	7	2	15%	120
B5	4	7	2.5	19%	120
B6	4	7	3	21%	120
B7	4	7	3.5	24%	120
B8	4	7	4	27%	120

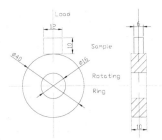

Figure 2. Schematic diagram of MM-200 experimental facility

RESULTS AND DISCUSSION

Friction coefficient

Figures 3 and 4 show the variation of friction coefficient with sliding distance at a load of 49N under dry and water lubricated conditions, respectively. As it can be seen, the friction coefficient of Si$_3$N$_4$-SiC and AlN-Si$_3$N$_4$-SiC ceramics materials is lower the friction coefficient of ceramics materials compared to Si$_3$N$_4$ ceramics irrespective of friction conditions. It can also be seen from Figs.2 and 3 that friction coefficient of Si$_3$N$_4$-based compound ceramic materials under water-lubricated condition is lower than that under dry friction condition, which indicates that Si$_3$N$_4$-based compound ceramics materials have better wear resistant under water-lubricated condition.

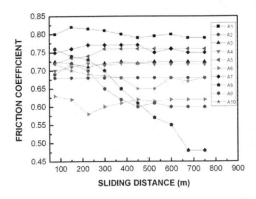

Figure 3. Variation of friction coefficient of reactive sintered Si$_3$N$_4$-based compound ceramics materials and Si$_3$N$_4$ ceramics with sliding distance under dry friction condition

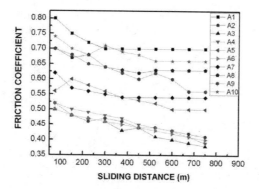

Figure 4. Variation of friction coefficient of reactive sintered Si₃N₄-based compound ceramics materials and Si₃N₄ ceramics with sliding distance under water-lubricated condition

Steady state friction coefficient

At the initial stage of friction, the variation of friction coefficient is large. When the abrasion enters into the steady period, the variation of friction coefficient tends to become steady, the friction coefficient at this period is defined as steady state friction coefficient of materials abrasion.

Under the condition of water-lubricated, when the load was 49N the variation of steady state friction coefficient of reactive sintered AlN-Si₃N₄-SiC with amount of Al addition is given in Fig.5. The steady state friction coefficient of Si₃N₄ ceramics material is initially observed to be 0.7. The results showed that all Si₃N₄-SiC ceramics and AlN-Si₃N₄-SiC ceramics can lower the friction coefficient of materials effectively. The lowest steady state friction coefficient of 0.38 is obtained when the amount of Al addition is 9%.

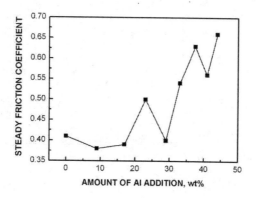

Figure 5. Variation of steady state friction coefficient with addition amount of Al

The variation of steady state friction coefficient with SiC content in reactive sintered AlN-Si₃N₄-SiC under the condition of water-lubrication at a load of 49N is given in Fig.6. The results show that the steady state friction coefficient decreases with the increasing amount of SiC addition. The lowest steady state friction coefficient of 0.39 when amount of Al addition is 27% indicates that friction performance of compound materials can be greatly improved with the increase of amount of SiC addition when the addition proportion of Al and Si is fixed in materials.

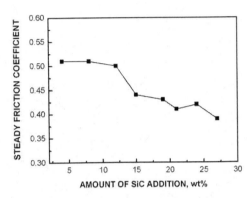

Figure 6. Variation of steady state friction coefficient with amount of SiC addition

Wear rate

Under the condition of water-lubrication and the load of 49N, when the reactively sintered Si₃N₄-based compound ceramics contacts with Si₃N₄ ceramics ring, the variation of the wear rate of reactive sintered Si₃N₄-based compound ceramics with amount of Al addition is given in Fig.7. The wear rate of Si₃N₄ ceramics material is found to be $1 \times 10^{-4} mm^3 (N \cdot m)^{-1}$ under the same friction condition. The results reveal that Si₃N₄-SiC ceramics and AlN-Si₃N₄-SiC ceramics can lower the wear rate of Si₃N₄ ceramics greatly. As it can be seen, the wear rate of the compound ceramics materials is the lowest when amount of Al addition is 37.5%, the wear rate of the compound ceramics materials is almost the same to that of unmodified Si₃N₄-SiC when the amount of Al addition is lower than 17%. As it can also be seen, addition of Al can not improve friction performance effectively and the wear rate of the compound ceramic materials is higher than Si₃N₄-SiC when amount of Al addition is observed to be higher than 17% (except 37.5%). In other words, AlN-Si₃N₄-SiC ceramics is not absolutely superior to Si₃N₄-SiC with respect to improving the wear resistant of ceramics materials. At the same time, the wear resistance of compound ceramics is not effectively improved with the increases of additional amount of Al.

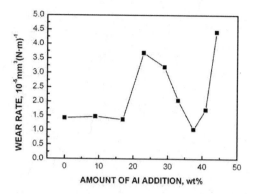

Figure 7. Variation of wear rate with amount of Al addition

Under the condition of water-lubricated, when the load was 49N and friction occurs between reactive sintered Si$_3$N$_4$-based compound ceramics and Si$_3$N$_4$ ceramics ring, the variation of the wear rate of reactive sintered Si$_3$N$_4$-based compound ceramics with amount of SiC addition was given in Fig.8. It showed that the wear rate of compound materials had a trend of fluctuations with the increase of Al additional amount, the wear rate of compound materials is the lowest and the wear resistance is the best when amount of SiC addition was 15%. No linear relation existed between wear rate and the amount of SiC addition because the factors influencing on the wear resistance of ceramics materials are multifarious.

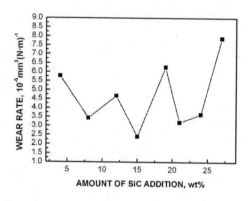

Figure 8. Variation of wear rate with amount of SiC addition

Load

Variation of wear rate of reactive sintered AlN-Si₃N₄-SiC compound ceramics materials with different loads is shown in Fig.9. As it can be seen, the higher the load, the higher wear rate of reactive sintered AlN-Si₃N₄-SiC compound ceramics materials. With the increase of load, temperature of friction surface of materials increases sharply and a large number of solid particles begin to spall. As a result, wear mechanism transforms from plastic deformation to brittle fracture, which further enhances the wear process and wear rate increases accordingly.

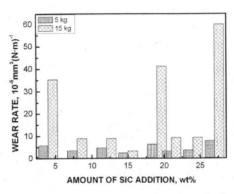

Figure 9. Variation of wear rate with different loads

Friction rings

GCr15 is a kind of high-carbon-chromium bearing steel which has higher hardness, even structure and good abrasive resistance. It is often used to estimate the wear-resisting property of materials. Variation of the steady friction coefficient of compound ceramics materials sliding against GCr15 ring and Si₃N₄ ring is given in Fig.10. It can be seen that the steady friction coefficient of ceramic samples sliding against GCr15 ring is lower than samples sliding against Si₃N₄ ring under the same friction condition, mean steady friction coefficient of the former is only 0.44. Therefore, friction performance of Si₃N₄-based compound ceramics materials sliding against metal is observed to be superior to that sliding against ceramics obviously.

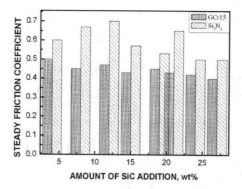

Figure 10. Variation of the steady friction coefficient of compound ceramics materials sliding against GCr15 ring and Si₃N₄ ring

Under the condition of water-lubrication, when the load is 49N and reactive sintered Si_3N_4-based compound ceramics materials slide against GCr15 ring, variation of friction coefficient of compound ceramics with slip distance is given in Fig.11. As it can be seen, friction coefficient of compound ceramics increases with the increase of slip distance and tends to stabilize after some time, this is because asperities between surface of steel ring and surface of ceramics samples drop gradually and form the tiny wear particles with the proceeding of the friction. With the continuing process of friction, the asperities of wear surface disappear and wear particles adhere to materials surface. As a result, the abrasion process enters into steady period and the friction coefficient tends towards a steady value.

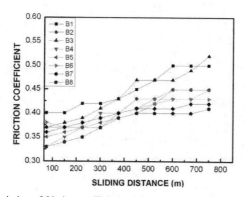

Figure 11. Variation of friction coefficient of compound ceramics with sliding distance

Variation of wear rate of reactive sintered Si_3N_4-based ceramics compound materials sliding against Si_3N_4 ring and GCr15 ring with amount of SiC addition is given in Fig.12. It can be seen that the wear rate of ceramic samples sliding against GCr15 ring is lower than samples sliding against Si₃N₄ ring under the same friction condition which is even two orders of magnitude lower.

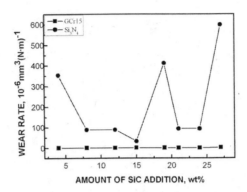

Figure 12. Variation of wear rate of reactive sintered Si₃N₄-based ceramics compound materials sliding against Si₃N₄ ring and GCr15 ring with amount of SiC addition

Zhao [4] reported that good cooling effect of the water has a peculiar effect on lowering the energy of friction surface under the condition of dry friction and water- lubrication. The presence of products like FeSiO₃ and FeO etc. on the worn surface as revealed by EDS analysis from Fig.13 indicates the occurrence of a chemical reaction on the worn surface during the friction. Generation of product FeSiO₃ and FeO etc can improve the boundary lubrication state of friction surface and thus reduce the wear rate of Si₃N₄ ceramics.

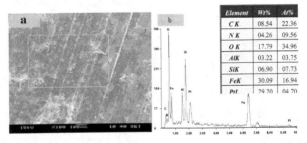

Element	Wt%	At%
C K	08.54	22.36
N K	04.26	09.56
O K	17.79	34.96
AlK	03.22	03.75
SiK	06.90	07.73
FeK	30.09	16.94
PtL	29.20	04.70

Figure 13. X-ray EDS Analysis of abrasion surface of B4 sliding against GCr15 ring under water-lubricated

CONCLUSIONS

Based on the results of this study, the following conclusions may be drawn:

1. The friction coefficient of Si₃N₄-SiC and AlN-Si₃N₄-SiC ceramics is lower compared to Si₃N₄ ceramics. The steady friction coefficient of compound materials is increased with the increase of the addition amount of Al and decreased with the increase of the addition amount of SiC.

2. The wear rate of Si$_3$N$_4$-SiC and AlN-Si$_3$N$_4$-SiC ceramics is lower compared to Si$_3$N$_4$ ceramics under the load of 49N and water-lubrication.
3. The frictional and wear properties of materials is greatly influenced by load. Both the friction coefficient and the wear rate increase with the increase of load under the condition of water-lubrication.
4. Frictional and wear behaviors of ceramic materials are greatly different when sliding against different counterface. The wear rate of ceramic samples sliding against GCr15 ring is found to be two orders of magnitude lower than that of samples sliding against Si$_3$N$_4$ ring under the same friction condition.

ACKNOWLEDGEMENTS

The work described in this paper was supported by Scientific Research Program of Jiangsu economic information committee, Program for Postgraduate Research Innovation in University of Jiangsu Province (CX10B_122Z) and Science and Technology Project of Jiangsu Province Construction System (JS2010JH22).

REFERENCES

[1] YUAN Jin. PREPARATION OF REACTIVE SINTERED Si$_3$N$_4$-BASED COMPOSITE CERAMIC. Nanjing University of Science and Technology master degree thesis. 2004.
[2] CUI Chong, WANG Yuan-ting, Preparation of AlN-Si$_3$N$_4$-SiC Composite Ceramic Materials,CN 200510041261. 9 (2007).
[3] CUI Chong, WANG Yuan-ting, and JIANG Jin-guo, Microstructure of reactive sintered Al bonded Si$_3$N$_4$-SiC ceramics, Trans. Nonferrous Met. Soc. China, 16, s42-s45 (2006).
[4] Zhao Xing zhong, Liu Jia jun, and Zhu Bao liang, TRIBOLOGICAL PROPERTIES OF Si$_3$N$_4$ / 1045 STEEL PAIRS UNDER LUBRICATED AND UNLUBRICATED CONDITIONS, Journal of the Chinese Ceramic Society, 10, 515~521 (1996).

PREPARATION OF NANO AND MICRON SIZED ZrO$_2$ DISPERSED Al$_2$O$_3$ CERAMIC COMPOSITES AND STUDY THEIR HARDNESS AND FRACTURE TOUGHNESSES

Kuntal Maiti, Anjan Sil
Department of Metallurgical and Materials Engineering
Indian Institute of Technology Roorkee, Roorkee - 247667, Uttarakhand, India

ABSTRACT

The present study aims at improving the fracture toughness of bulk alumina ceramics by incorporating micron and nano sized zirconia particles. Al$_2$O$_3$ – ZrO$_2$ ceramic composites were prepared with 1vol% micron sized and 1vol% nano sized ZrO$_2$ particles by uniaxial pressing and sintering at 1500°C for soaking time period of 3hr, 6hr and 9hr. The mechanical properties viz. hardness and fracture toughness of the composites were investigated. It was found that the fracture toughness of the nano/micron sized zirconia particle dispersed Al$_2$O$_3$ composites was significantly improved compared to monolithic Al$_2$O$_3$. The alumina composites with 1vol% nano sized ZrO$_2$ particles sintered at 1500°C for 6hr show highest fracture toughness of 5.76 MPam$^{1/2}$, while the maximum fracture toughnesses of only alumina and alumina composites with 1vol% micron sized ZrO$_2$ particles are 5.29 and 5.35 MPam$^{1/2}$ respectively. The toughening mechanisms of the ceramic composites are discussed.

INTRODUCTION

The field of ceramics is vast and varied. Al$_2$O$_3$ is considered a typical representative of the engineering ceramics. The properties of the engineering ceramics are particularly important for structural applications such as in the motor, aerospace and biomedical fields, especially when the environmental conditions are particularly severe[1]. The brittleness and poor damage tolerance have, so far, limited the application of structural ceramics as advanced engineering materials[2]. The fracture toughness of the materials is low because of dislocation movement is extremely limited by the type of bonds which are ionic and/or covalent. The problem of the low fracture toughness of ceramics can be overcome by designing and preparing composite materials reinforced with fibres, whiskers and particulates[3, 4, 5]. The addition of one or more components into the base material to form ceramic matrix nano composites (CMNCs) has been found to be effective to enhance the fracture toughness and strength of the materials[6]. The main objective of the present work is to investigate the effects of nano and micron sized ZrO$_2$ particle dispersion in Al$_2$O$_3$ matrix on hardness and fracture toughness in Al$_2$O$_3$ – ZrO$_2$ composites. The ceramic composites were prepared and their microstructural observations were made. The density, linear shrinkage and weight loss of the composites after sintering have been estimated and reported. The hardness and fracture toughness of these composites were determined at room temperature.

EXPERIMENTAL PROCEDURE

The raw materials used are Al$_2$O$_3$ (Industrial Ceramics, 99.8%, 0.4µm), micron sized ZrO$_2$ particles (Himedia, 0.4µm) and nano sized ZrO$_2$ particles (American Elements, 40nm). Al$_2$O$_3$ composites were prepared by dispersing 1vol% of micron sized ZrO$_2$ powder and 1vol% of nano sized ZrO$_2$ powder separately into Al$_2$O$_3$ matrix. ZrO$_2$ was added into distilled water and ultrasonicater was used to deagglomerate the nano powders. After the sonication treatment for 6hr the deagglomerated powder mixing was carried out by using magnetic stirrer for 36hrs. The mixture was dried at 100°C in an oven. The dried mixture was lightly crushed for deagglomeration, ground in a mortar and pestle. Granulation was done with the addition of binder polyvinyl alcohol.

The granulated powder was compacted uniaxially in a hydraulic press at different loads ranging between 15×10^4 and 22.5×10^4 N, in the form of rectangular bars having dimension of 3mm x 5mm x 45mm. The samples were sintered at 1500°C for different soaking periods of 3hr, 6hr and 9hr. The samples were heated up to 1200°C at a heating rate of 10°C/min and between 1200°C and 1500°C the heating rate was kept at 5°C/min. The cooling upto the room temperature was set at the rate of 10°C/min. However, the samples were actually furnace cooled. The volume shrinkage and linear shrinkage of the samples after sintering were measured by digital vernier caliper. The density of all the samples was measured following Archimedes principle using water as an immersion medium. The relative densities of the samples were estimated based on the theoretical densities calculated following the rule of mixture (ROM) for ceramic composites. The samples were ground on one face using silicon carbide (SiC) papers of # 120, # 220, # 320 grades to develop smooth surface finish. The smooth surfaces of the samples were polished with diamond pastes of 6 and 3μm sizes, to the level of mirror like finish, and indented using Vickers indenter with a load of 5 kg for a dwell time of 15s under ambient condition. The diagonals of the square shaped indentation and crack lengths generated largely from corners of the indentation were measured by optical microscope (Axiovert 200 MAT, ZEISS). The hardness (H_v) of the samples was calculated from the following equation:

$$H_v = P/A = \alpha P / d_o^2 \tag{1}$$

where P is the applied load, A is the pyramidal contact area of the indentation, d_o is the average length of the diagonals of the resultant impression and $\alpha = 1.8544$ for Vickers indenter.

Fracture toughness of the rectangular samples was determined by Single Edge Precracked Beam (SEPB) method with 20mm span and the cross head speed of 0.5mm/min.

The microstructures of the samples were observed by FESEM (FEI, QUANTA 200F). The polished sample surface was gold coated for the SEM observations. The grain size of the sintered samples was measured from the micrographs by using Image J software.

The following nomenclatures A, AMZ and ANZ have been used to represent the materials of Al_2O_3 and composites Al_2O_3 – 1vol% micron sized ZrO_2, Al_2O_3 – 1vol% nano sized ZrO_2 respectively.

RESULTS AND DISCUSSION

Table I shows that the sintered densities of ANZ samples sintered for all the three time periods are consistent compared to the A and AMZ samples. The density of the ANZ samples on the whole is higher than those of he samples A and AMZ sintered for all the soaking time periods. This can be attributed to the higher packing density resulted in the green compact of ANZ due to nano sized ZrO_2 particles, which occupy part of the voids among Al_2O_3 particles [6]. Linear shrinkages of the samples are nearly equal in all three A, AMZ and ANZ systems. The weight loss in ANZ and AMZ samples is marginally less than that of A. The weight loss of the samples has been investigated just to strengthen the physical characterization of the samples. The more weight loss implies the presence of higher content of volatile materials in the green samples. The removal of the volatile content during sintering the samples leads to the porosity development in the sample. The weight loss in the cases of ANZ and AMZ samples are less compared to the A samples.

Table I. Density, Shrinkage and Average Grain Size of A, AMZ and ANZ Sintered at 1500°C for 3 hr, 6hr and 9hr.

Sample	Soaking Time (hr)	Sintered Density (g/cm³)	Relative Density (%)	Linear Shrinkage (%)	Weight Loss (%)	Average Grain Size (μm)
A	3	3.83	96.23	14.19	6.42	0.59
	6	3.90	97.99	14.25	6.44	0.83
	9	3.92	98.49	14.54	6.53	0.85
AMZ	3	3.87	96.75	13.91	6.07	0.69
	6	3.90	97.50	13.85	6.02	1.19
	9	3.89	97.25	14.07	6.09	1.27
ANZ	3	3.90	97.50	14.16	6.04	0.81
	6	3.92	97.99	13.95	6.11	1.32
	9	3.91	98.25	14.25	6.14	1.40

MICROSTRUCTURAL ANALYSIS

Effect of micron and nano sized ZrO_2 powder dispersion on microstructure of Al_2O_3 - ZrO_2 samples

Figure 1 shows that the average grain size is higher in AMZ and ANZ samples compared to that in the A sample. The grain size data are given in Table I. The average grain size in ANZ is maximum. The ZrO_2 phase is shown by arrows in the micrographs (Figure 1). Microstucture in Figure 1(b) reveals that micron sized ZrO_2 is present as separated particle as well as in the form of agglomeration, at the grain boundaries and grain junctions. In ANZ the nano ZrO_2 particles, are largely situated at the grain boundaries and within the grains also. Though the grains are not seen (from the micrographs) clearly separated by distinct grain boundaries, but the observations on grain sizes have been made for the estimation of average grain size.

Figure 1. SEM micrograph of (a) A, (b) AMZ and (c) ANZ composites sintered at 1500°C for 6hr.

Grain size histogram of Al$_2$O$_3$ and Al$_2$O$_3$- ZrO$_2$

As the grains in all the samples vary over a range of size rather than preferably having uniform size for a particular sintering schedule, it is appropriate to show the actual size distribution that has been resulted due to sintering. The measured grain sizes were subdivided into small size intervals and plots were made in the form of histograms to show the grain size distribution. Figure 2 presents such histograms for the samples sintered at A$_6^{1500}$, AMZ$_6^{1500}$ and ANZ$_6^{1500}$. It can be seen from the Figure 2(a) that though the grains in the sample A are of different sizes varying from about 0.3 to about 1.9μm, a large number of grains fall in the lower size subrange of 0.4 - 0.6μm. The grain size ranges of the samples AMZ and ANZ as shown in Figure 2(b) and (c) are larger than that of the sample A. The size ranges are 0.3 – 4.5μm and 0.3 – 6.0μm respectively for AMZ and ANZ.

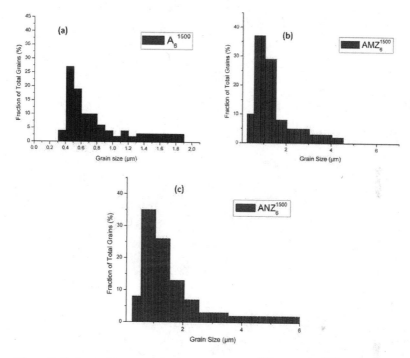

Figure 2. Grain size histograms for the samples (a) A_6^{1500}, (b) AMZ_6^{1500} and (c) ANZ_6^{1500}.

Thermogravimetric analysis of powdered A sample has been conducted and the TG profile is plotted in Figure 3. It can be seen from the TG plot that the weight loss of about 2.5% occurs during heating the raw Al_2O_3 powder sample upto 1500°C. The remaining weight loss (Table I) due to sintering of the samples can be accounted for binder and solvent removal[7].

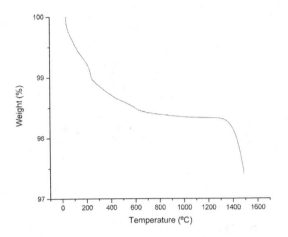

Figure 3. TG curve of Al_2O_3 powder

MECHANICAL PROPERTIES

Table II. Fracture toughness and hardness of A, AMZ and ANZ samples sintered at 1500°C for 3hr ,6hr and 9hr.

Sample	A			AMZ			ANZ		
Soaking Time Period (hr)	3	6	9	3	6	9	3	6	9
Fracture Toughness $(MPam^{1/2})$	4.66	5.26	5.29	5.05	5.35	5.28	5.54	5.76	5.24
Hardness (GPa)	15.01	14.94	14.77	14.82	14.68	14.73	14.95	14.93	14.79

Table II shows the mechanical properties of the samples A, AMZ and ANZ sintered at 1500°C for different soaking time periods. The hardness of all types of samples A, AMZ and ANZ decreases marginally with the increase in soaking period. Such gradual lowering of the hardness is attributed to the increase in grain size with the increase in soaking time for each type of the samples[8]. The range of fracture toughness of the sample ANZ is 5.24 – 5.76 MPam$^{1/2}$ which is higher compared to the sample AMZ and A for which the ranges are 5.05 – 5.35 MPam$^{1/2}$ and 4.66 – 5.29 MPam$^{1/2}$ respectively.

Microstructure of fracture surface

Figure 4. SEM micrographs of fracture surfaces of (a) A, (b) AMZ and (c) ANZ composites sintered at 1500°C for 6 hr.

Figure 4(a) shows the fracture surface of A sample in which the fracture has taken place predominantly by the intergranular mode. Figure 4(b) also shows predominantly intergranular fracture mode but the presence of the ZrO_2 particles in the triple point is evident. Thermal expansion coefficient mismatch between Al_2O_3 and ZrO_2 creates thermal stress. The thermal stress increases the fracture toughness by deflecting the crack path[9, 10]. Figure 4(c) shows the fracture surface of ANZ where fracture has taken place due to mixed mode of intergranular and transgranular fracture[11]. The fracture toughnesses of these samples (Figure 4(c)) show relatively higher toughness as listed in Table II.

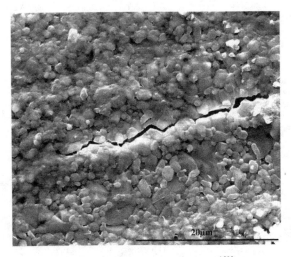

Figure 5. SEM micrograph of ANZ_6^{1500}.

Figure 5 shows a FESEM micrograph of the ANZ_6^{1500} sample surface showing an enlarged view of the crack emanating from the corners of the Vickers indentation on the sample. The zig-zag nature of the crack path as can be seen in the micrograph is an indicative of the crack deflection mode being operative. Therefore the crack deflection is the predominant mechanism for toughness enhancement in this sample.

CONCLUSION

The following conclusions can be drawn from the present study:

1. Though the grain size range in all the three types of samples A, AMZ and ANZ is wider, however, a major fraction of the grains fall in the lower part of the respective size ranges. Thus a significant portion of the grains in AMZ and ANZ samples is of around 1μm, whereas in A sample the size is around 0.6μm.

2. Hardness of the A, AMZ and ANZ samples measured by Vickers indentation technique lies in the range of 14.68 to 15.01 GPa.

3. The fracture toughness of the ANZ is found generally higher than that of AMZ and A. The highest fracture toughness of 5.76 MPa m$^{1/2}$ is found with the ANZ_6^{1500}.

4. The fracture in the ANZ is caused by mixed mode of intergranular and transgranular types.

REFERENCES

1. F. Cesari, L. Esposito, F.M. Furgiuele , C. Maletta, A. Tucci, Fracture Toughness of Alumina–Zirconia Composites, *Ceram. Int.*, **32**, 249 - 50 (2006).

2. Y.L.Dong, F.M.Xu, X.L.Shi, C.Zhang, J.M.Yang, Y.Tan, Fabrication and Mechanical Properties of Nano-/Micro-Sized Al_2O_3 / SiC Composites, *Mater. Sci. Eng. A*, **504**, 49- 54 (2009).

3. D. L. Porter, A. G. Evans and A. H. Heuer, Transformation-Toughening in Partially-Stabilized Zirconia (PSZ), *Acta Metall.*, **27**,1649-54 (1979).

4. S. Shukla, S. Seal, R. Vij, S. Bandyopadhyay and Z. Rahman, Effect of Nanocrystallite Morphology on the Metastable Tetragonal Phase Stabilization in Zirconia *Nano Latters*, **2**,989 - 93(2002).

5. T. Kosmac, M.V. Swain, N.Claussen, The Role of Tetragonal and Monoclinic ZrO_2 Particles in the Fracture Toughness of Al_2O_3-ZrO_2 Composites, *Mater. Sci. Eng. A*, **71**, 57-64 (1985).

6. F. A.T. Guimaraes , K. L. Silva, V. Trombini , J. J. Pierri, J. A. Rodrigues , R. Tomasi , E. M.J.A. Pallone, Correlation Between Microstructure and Mechanical Properties of Al_2O_3 /ZrO_2 Nanocomposites, *Ceram.Int.*, **35**, 741-45 (2009).

7. S.K. Durrani, J. Akhtar, M. Ahmad, M.A.Hussain, Synthesis and Characterization of Low Density Calcia Stabilized Zirconia Ceramic for High Temperature Furnace Application, Mater. Chem.Phy., **100**, 324 – 28 (2006).

8. H.X.Lu, H.W. Sun, G.X. Li, C.P.Chen, D.L. Yang, X.Hu, Microstructure and Mechanical Properties of Al_2O_3 –MgB_2 composites, *Ceram.Int.*, **31**, 105 - 08 (2005).

9. T. Ekstrom, Alumina Ceramics with Particle Inclusions, *J.Eur.Ceram.Soc.*, **11**, 487 - 96 (1993).

10. A.Nakahira, K. Niihara, Sintering Behaviors and Consolidation Process for Al_2O_3/SiC Nanocomposites *J. Ceram. Soc. Jpn.*, **100**, 448 - 53 (1992).

11. S.N. Chou , H.H.Lu, D.F. Lii, J.L. Huang, Investigation of Residual Stress Effects in an Alloy Reinforced Ceramic/Metal Composite, *J. Alloys Comp.*, **470**,117-22 (2009)

INFLUENCE OF PARTICLE SIZE DISTRIBUTION OF WOLLASTONITE ON THE MECHANICAL PROPERTIES OF CBPCs (CHEMICALLY BONDED PHOSPHATE CERAMICS)

H. A. Colorado[a,b,*], C. Hiel[c,d], H. T. Hahn[a,e,f]

[a]Materials Science and Engineering Department, University of California, Los Angeles, CA 90095, USA Email: hcoloradolopera@ucla.edu
[b]Universidad de Antioquia, Mechanical Engineering. Medellin-Colombia
[c]Composite Support and Solutions Inc. San Pedro, California
[d]MEMC-University of Brussels (VUB)
[e]Mechanical and Aerospace Engineering Department, University of California, Los Angeles
[f]Korea Institute of Science and Technology, KIST. Seoul, South Korea

ABSTRACT

This paper investigates the compressive strength and curing of chemically bonded phosphate ceramics (CBPC) as effected by the particle size of its Wollastonite ($CaSiO_3$) powder component. The CBPC was fabricated by mixing a patented aqueous phosphoric acid formulation with Wollastonite powder. The phosphoric acid formulation was adapted to the powder particle size in order to increase the setting time of the reaction. Optical and Scanning Electron Microscopy (SEM) were used to investigate the crystalline and amorphous phases, as well as the microstructure evolution. Additionally X-ray diffraction (XRD) was also utilized. The stability of the various phases with temperature was studied with Thermo Gravimetric Analysis (TGA).

INTRODUCTION

CBCs have been extensively used for multiple applications. They include: dental materials[1], bone tissue engineering[2], shielding gamma and neutron radiation[3], nuclear waste solidification and encapsulation[4], electronic materials[5], tooling for advanced composites[6], nanotubes and nanowires applications[7], high temperature[8], composites with fillers and reinforcements[9 and 10]. The Chemically Bonded Phosphate Ceramics can reach a compressive strength of 100MPa in minutes, whereas Portland cement based concrete reaches a compressive strength about 20MPa after 28 days. This opens up many applications related to infrastructure repair such as roads, bridges and pipes.

Chemically Bonded Ceramics (CBCs) are inorganic solids consolidated by chemical reactions at low temperatures without the use of thermally activated solid-state diffusion (typically less than 300°C), instead of high temperature processing (by thermal diffusion or melting) as is normally done in traditional ceramics[11, 12, 13 and 14].

The bonding in CBCs is a mixture of ionic, covalent, and van der Waals bonding, with the ionic and covalent dominating; unlike in traditional cement hydration products, in which van der Waals and hydrogen bonding dominate[15]. The chemical bonding in CBC's allows them to be inexpensive in high volume production. Therefore CBCs are environmentally benign and fill the gap between cements and ceramics. The fabrication of conventional cements and ceramics involve energy consuming high temperature processes, which adversely affect the environment.

The CBCs form by acid-base reactions between an acid phosphate and a metal oxide [15 and 10]. In the CBPCs, when a polyacid solution (in this research is a patented aqueous phosphoric acid formulation) and the metal oxide (in this research is Wollastonite, $CaSiO_3$) powder mixture are stirred, the sparsely alkaline oxides dissolve and an acid base reaction is initiated. The result is slurry that hardens in a ceramic product[14]. The setting is the result of gelation by salt formation and the cations (Ca^{+2} in this research) are extracted from the calcium phosphate[10 and 16]. Thus, we have two main process involved in the process, the release of cations from the Wollastonite and the interaction with the acidic solution. Thus, two reaction rates are competing: the rate of release of cations and the rate of the structure formation. If the rate of release of cations is too fast, a non-coherent precipitates of crystallites is formed. On the other hand, if is too slow, a gel formed with lack of strength[16]. A complete overview of the principles of the wider field of acid-based cements[16] and the details of the models to explain the kinetics of formation have been written before [17 to 19].

In this research, we used a CBC formed by Wollastonite powder and phosphoric acid (H_3PO_4), which, when mixed in a ratio of 100/120 reacts into a Chemically Bonded Phosphate Ceramic (CBPC). Wollastonite is a natural calcium meta-silicate which is mostly used as filler in resins and plastics, ceramics, metallurgy, biomaterials and other industrial applications[20].

The mixing of Wollastonite with phosphoric acid produces calcium phosphates (brushite ($CaHPO_4 \cdot 2H_2O$), monetite ($CaHPO_4$) and calcium dihydrogenphosphate monohydrate ($Ca(H_2PO_4)2 \cdot H_2O$)) and silica for molar ratios (P/Ca) between 1 and 1.66[21]. We found that a ratio of phosphoric acid to Wollastonite of 1.2 leads into a neutral product, with a pH of 7.0. This is particularly interesting when they are compared with Portland cement for which pH is in excess of 12, which does not allow the use of E-glass fibers as a reinforcement.

The sections below will present the effects of particle size distribution on mechanical properties of CBPCs. Curing and compression tests were performed for CBPCs made with Wollastonite powder of different sizes. The mechanism of weight loss is presented. The mechanism in which the CBPCs naturally go into a micro porous material is discussed in this paper.

EXPERIMENTAL

CBPC manufacturing

The manufacturing of CBPC samples was conducted by mixing an aqueous phosphoric acid formulation (from Pastone USA) and natural Wollastonite powder (from Minera Nyco; see Table 1 and 2) in a 1.2 ratio liquid/powder. The pH of the CBPC after curing was 7.0.

Table 1 Chemical composition of Wollastonite powder.

Composition	CaO	SiO_2	Fe_2O_3	Al_2O_3	MnO	MgO	TiO_2	K_2O
Percentage	46.25	52.00	0.25	0.40	0.025	0.50	0.025	0.15

The median particle size was determined with a cilas granulometer, the surface area was determined with an ASAP 2405 (Micromeritics), and the moisture content was determined with a Karl Fischer instrument. These properties are presented in Table 2. In all cases pH (10% slurry) was 9.9.

Table 2 Properties of Wollastonite powders as received

Powder reference	Median particle size (μm)	Surface Area (m^2/g) (BET)	Moisture (%)
M200	15	1.1	0.05
M400	8	1.6	0.20
M1250	3.5	2.9	0.25

Curing and compression samples were fabricated. The mixing process of Wollastonite powder and phosphoric acid formulation was done in a Planetary Centrifugal Mixer (Thinky Mixer® AR-250, TM). For curing test samples, both the Wollastonite powder and the phosphoric acid formulation were maintained at room temperature. For compression samples, both the Wollastonite powder and the phosphoric acid formulation were maintained at 3 °C in a closed container (to prevent water absorption) for 1 hour in order to increase the pot life of the resin.

Curing tests

For all cure samples, 24g of the acid solution and 20g Wollastonite powder (both at room temperature) were mixed in the TM initially for 5 sec to release most of the upcoming gases, and then for another 5 sec to achieve a homogeneous mixture. Three factors were taken into account to select this short mixing time:

- some Wollastonite powder sizes react very fast;
- the small amount of precursor (44g) requires less mixing time; and
- at room temperature it is easier to see the effect on the setting time of different powders, additives and processing methods.

Curing curves were obtained measuring the temperature evolution with time for different powder sizes. Both mixing and curing were conducted in TM containers. The mixing was performed in 125ml and the curing in 24ml Polypropylene jars. A thermocouple was inserted through a hole in the cap of the 24ml jar, and sealed as shown in Figure 1a. The same containers (for mixing and then for curing) were used following the same parameters and environmental conditions for all samples. Figure 1b shows samples after the curing test as well as compression samples. Additional curing tests were conducted by first placing 100g of Wollastonite powder in the oven for 2h at 300°C in order to remove all water content. Samples were cooled to 200°C while remaining in the oven, (which took 1h) and subsequently cooled in ambient air to room temperature. They were kept in a container open to air (mixed mechanically twice per day to expose all particles surface) in order to see the humidity effect on curing of CBPCs.

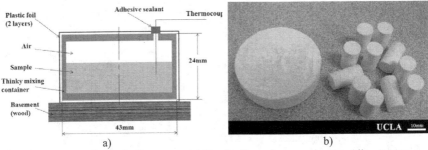

a)   b)

Figure 1 a) Experimental set up representation for the curing experiments. All tests were conducted at room temperature in a closed (to humidity) plastic container, with a thermocouple put it into a hole then sealed with tape; b) typical samples after the curing test (left) and before compression test (right).

Compression tests

For all compression samples, 120g of the acid solution and 100g of Wollastonite, both at 3 °C, were mixed in the TM initially for 10 sec to release most of the upcoming gases, then for 2 min. A Teflon® fluoropolymer mold with mold release (Synlube 1000 silicone-based release agent applied before the mixture discharge) was used to minimize the adhesion of the CBPC to the mold. Then, the mold with the CBPC was covered with plastic foil to preserve humidity and decrease the shrinkage effects. Samples were released after 48 hours and then dried at room temperature in open air for 3 days. For samples of this thickness it was found this amount of time was enough to stabilize almost all weight loss at temperatures near room temperature. Then, samples were mechanically polished with parallel and smooth faces (top and bottom) for the compression test. Since the CBPC has both unbonded and bonded water, a thermal treatment (to completely dry the samples) was conducted in order to stabilize the weight loss. Some samples were dried slowly in the furnace in order to prevent residual stresses. The thermal treatment sequence was: 50°C for 1 day, followed by 105°C for 1 day, followed by 205°C for 1 day. Examples of CBPC compression samples are presented in Figure 1b.

Compression tests were conducted in an Instron® machine 3382, over cylindrical CBPC samples (9mm in diameter by 20mm in length) for M200, M400 and M1250 Wollastonite powders with the same age (two years old). CBPC fabricated with one week old M200 Wollastonite powder was also included to see the effect of aging on the compression strength. A set of 5 samples were tested for each powder. The crosshead speed was 1mm/min.

Other Characterization:

The pH was measured with Whatman pH indicator paper 0-14. Solid samples were ground and then diluted in 160 mg of deionized water. The pH of the Wollastonite powders and for the phosphoric acid formulation was 9.9 (see Table 2) and 1.0 respectively; however, the pH for the CBPC was 7.0. For molar ratios between 1.0 and 1.2 it was found a neutral product.

To see the microstructure, sample sections were ground using silicon carbide papers of 500, 1000, 2400 and 4000 grit progressively. Then they were polished with alumina powders of 1, 0.3 and 0.05μm grain size progressively. After polishing, samples were dried in a furnace at 50°C for 24 hours and observed in an optical microscope.

Additional samples for SEM examination were mounted on an aluminum stub and sputtered in a Hummer 6.2 system (15mA AC for 30 sec) creating a 1nm thick film of Au. The SEM used was a JEOL JSM 6700R in a high vacuum mode. Elemental distribution x-ray maps were collected on the SEM equipped with an energy-dispersive analyzer (SEM-EDS). The images were collected on the polished and gold-coated samples, with a counting time of 51.2 ms/pixel.

X-Ray Diffraction (XRD) experiments were conducted usin X'Pert PRO equipment (Cu Kα radiation, λ=1.5406 Å), at 45KV and scanning between 10° and 80°. M200, M400 and M1250 Wollastonite samples (before and after the drying process) were ground in an alumina mortar and XRD tests were done at room temperature.

Thermo gravimetric Analysis (TGA) was performed in a Perkin Elmer Instruments Pyris Diamond TG/DTA equipment. CBPC samples made with M200, M400 and M1250 Wollastonite powders after the drying process were ground on an alumina mortar were the temperature ramp was 2°C/min. Experiments were conducted in an Ar atmosphere.

A linear shrinkage test was performed following the ASTM C326-09 standard for drying and firing shrinkages of ceramic whiteware clays. Test specimens were bars of 19mm in diameter by 127 mm in length. Shrinkage reference lines were made with a knife at zero and 102 mm. The drying process was performed at 50 °C for 24 hours, followed by 105 °C for 24 hours. In addition, an adaptation to the procedure was added drying at 200°C for 24 hours. After all annealing times, the samples were cooled in the oven until it reached room temperature.

Finally, density tests were conducted over CBPC samples made with M200, M400 and M1250 Wollastonite powders after a drying process (50°C for 1 day, followed by 105°C for 1 day, followed by 205°C for 1 day) in a Metter Toledo^{TM} balance, by means of the buoyancy method.

The main parameters were:
- Temperature of water (24.2˚C),
- Air density (0.0012g/cm^3),
- Water density (0.99727 g/cm^3 at 24.2 C) and
- Balance correction factor (0.99985).

The weight of the sample in air as well as the weight of the sample in water (after they were boiled in a separate container) were taken three times for each sample, then the density was well as the volume were calculated with the main values.

ANALYSIS AND RESULTS

Phases were identified in detail with SEM-EDAX and XRD. Figure 2 shows X-ray maps for the cross section of CBPCs, all over the same region of the secondary electron image (SEI) of Figure 2a. The composition distribution of calcium, silicon and phosphorous is shown in Figure 2b, c and d respectively. Carbon and oxygen distributions were also obtained but are not presented because they were almost constant over the entire image.

Figure 2b shows evidence of some Wollastonite ($CaSiO_3$) grains which transformed into silica (by giving up calcium) and some which did not transform. Figure 2c shows silica glass as well as Wollastonite grains. Figure 2d shows the calcium phosphate matrix. Numbers 1, 2 and 3 correspond to silica, Wollastonite and calcium phosphates respectively.

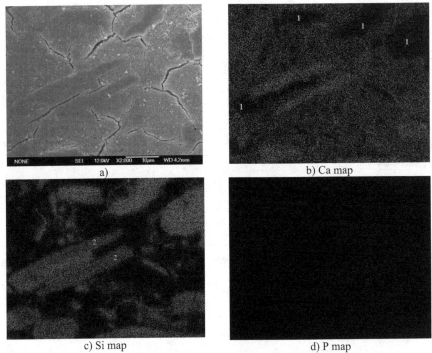

Figure 2 X-ray maps of pultruded CBPC, a) topographical image; and b) Ca, c) Si and d) P concentration images. Numbers 1, 2 and 3 correspond to silica, Wollastonite and calcium phosphates respectively.

Figure 3a shows a general SEM image of the CBPC with microcracks. Microvoids (not detected in this characterization) are really rare because the technique (TM) used effectively for mixing eliminates them (Resonant Acoustic Mixing has also proven as effective). Smooth zones correspond to silica and Wollastonite (partially or un-reacted) grains.

Figure 3b is a magnification of the center of Figure 3a showing that the matrix is composed of small crystals of calcium phosphate (brushite) and amorphous calcium phosphate. In some of these silica and Wollastonite grains, interfacial cracks appear. However, most of the cracks grow and propagate in the calcium phosphate matrix until they are eventually stopped by the grains. The silica and remaining Wollastonite particles provide nuclei for crystallization and can act as a second phase that reinforces the matrix[10]. As the drying process occurs, shrinkage and micro crack growth increase. As a result, interfacial cracks appear in grain interfaces between both silica and Wollastonite grains with the calcium phosphate matrix. A representation of the phenomena is presented in Figure 3c. The drying process is natural because of the excess water present in the aqueous phosphoric acid formulation. Cracking is caused by water evaporation induced shrinkage from the calcium phosphates, competing with the non-shrink silica and Wollastonite phases. As shrinkage increases in the calcium phosphate, the stress increases in the silica and Wollastonite phases. Since all phases in the CBPCs microstructure are fragile, micro cracking occurs.

In addition, the phases have dissimilar microstructure, which decreases interfacial strength. Silica is amorphous, Wollastonite is triclinic, calcium phosphates can be monoclinic (brushite) or amorphous and therefore poor interface strength is to be expected.

It has been reported[16] that if the rate of cation-release is too fast, non-coherent precipitates of crystallites are formed. However, for the CBPC fabricated in this research, it has been explained that the structural shrinkage driven by water release is the major reason for interfacial cracking.

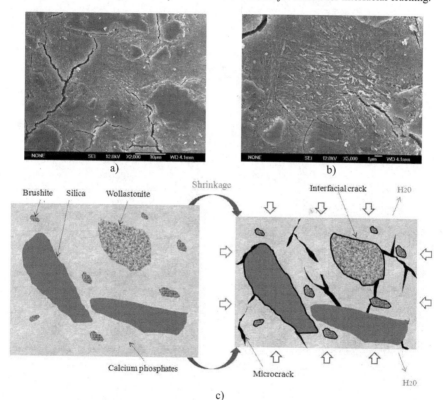

c)

Figure 3 a) CBPC SEM image made with M200 Wollastonite powder, b) a higher magnification at the center of the previous image, c) representation of the CBPC before and after the shrinkage process starts.

Results from the linear longitudinal shrinkage test following the ASTM C326-09 standard are presented in Figure 4. Samples tested at 105 °C and 200 °C had macro cracks and some residual bending deformation. Figure 4 also illustrates that the longitudinal shrinkage is lower than the diametral shrinkage.

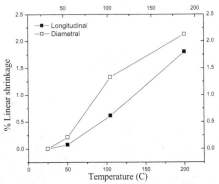

Figure 4 Drying and firing shrinkage test of CBPC made with M200 Wollastonite powder two years old following the ASTM C326-09 for linear longitudinal test. In addition, diameter shrinkage was also obtained.

Figure 5a shows curing curves of CBPCs made with M200, M400 and M1250 Wollastonite powders, all at room temperature. For the curing curve of the CBPC made with M200 Wollastonite powder, the higher temperature values at the shorter times correspond to the heat generated for the mixing in the TM; the other powders do not show this effect because they react faster and get hotter than M200. The size effect on the curing is shown in Figure 5b, from peaks presented in Figure 5a. An exponential decay first order fit was performed.

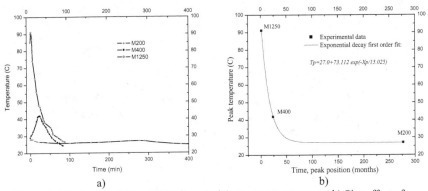

a)   b)

Figure 5 a) Curing for different CBPCs materials at room temperature, b) Size effect of Wollastonite powder on the CBPC setting taken from a).

Aging of Wollastonite powder has also been examined. The reactivity of the powder (observed when is mixed with the aqueous phosphoric acid formulation) decreased with age even though samples were kept in a sealed plastic container. This is caused by moisture absorption (Wollastonite is a hygroscopic material). Figure 6a shows the aging effects of M200

Wollastonite powder in the curing of CBPCs. The difference between the two years and one week old samples is striking: the two years old sample has an increase in the peak position (time) around four times the corresponding one for the one week old sample.

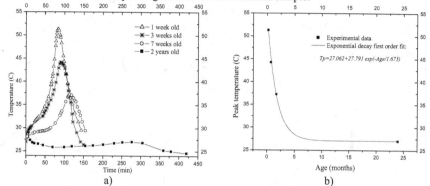

Figure 6 a) curing curves for CBPC made with M200 aged at different times; b) exponential decay first order fit for peak positions points obtained from a).

Figure 6b shows the aging effect on M200 Wollastonite powder. The peak temperature decreases and the time associated with these peaks increases as the powder gets older. The reactivity decreases because the ambient humidity from air reduces the calcium action on the reaction.

Finally, curing tests for CBPC made with M1250 Wollastonite powder after expose to 300C for 2h in air atmosphere are presented in Figure 7. This phenomenon can be associated with the known behavior of Wollastonite in contact with water, in which there is[20] a rapid release of surface Ca^{2+} replacement by $2H^+$. Thus, under the air moisture, the Wollastonite surface becomes more stable which decreases its reactivity.

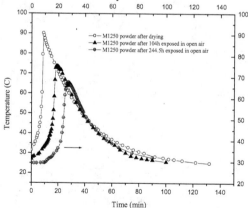

Figure 7 curing curves for CBPC made with M1250 after exposed to 300C for 2 hours and then put it in an open container open to air.

The compressive strength for the CBPCs made with two years old M200, M400 and M1250 Wollastonite powders as well as CBPC fabricated with one week old M200 Wollastonite powder are presented in Figure 8.

The error bars are slightly higher for the small powders sizes, which can be caused by increased porosity due to the high reactivity. As expected, the highest values (not the mean) are observed for the small sizes. The manufacturing problem is minimized by cooling the raw materials at temperatures of around 3°C as was done in this research; however, indications in Figure 8 are that the M400 particle size distribution creates a maximum compressive strength. On the other hand, the compression strength for two year old CBPC made with M200 is 45% less than for the baseline material made with two week old M200. This effect can also be associated with the moisture effect on the Wollastonite surface, which decreases its reactivity when it is mixed with the aqueous phosphoric acid formulation.

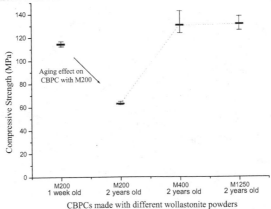

Figure 8 Compressive strength for CBPCs made with M200, M400 and M1250 Wollastonite powders. The green dot line indicates powders with same age.

Density tests were performed over the two years old samples tested in compression for CBPCs made with M200, M400 and M1250 after the following drying process: 50°C for 1 day, followed by 105°C for 1 day, followed by 205°C for 1 day. Results, presented in Table 3, show that as the mean of particle distribution decrease (from M200 to M1250), the density decreases. This result can have two contributions:

- As the powder size decreases, more surface area allows for a more complete reaction (less unreacted Wollastonite particles in the CBPCs) which means more calcium phosphate, which is the more porous, micro-crack containing phase of the CBPC.
- Since reactivity is very fast for small powder sizes, it is more likely that internal pores be introduced in the mold since the paste was very viscous. This explains the higher variability in compression for CBPCs made with the small powders as compared to CBPC made with M200

On the other hand, the density mean value of 2.19 g/cm³ is particularly interesting when compared with Portland cement, for which is 3.15 g/cm³. Even lower density values have been reported for other CBPCs[10].

Table 3 Density values for different CBPCs after a drying process

Powder reference	M200	M400	M1250	Mean
Density (g/cm³)	2.319	2.146	2.112	2.19
Volume (cm³)	0.904	0.903	0.900	0.91

XRD characterization taken for different Wollastonite powders under different processing conditions (Figure 9a) did not show significant changes. The results presented above for curing show that the bulk powder structure does not change significantly with air exposure or thermal treatments below 300C. However, for CBPCs, other properties like curing time and temperature peak of the reaction are affected considerably. As a consequence, setting time, viscosity and mechanical properties can be notably changed. These results indicate that we have a process where the Wollastonite surface has a very significant effect on the CBPC material properties.

XRD for CBPCs (Figure 9b) showed very small changes when different powders were used. In general, M1250, the smallest powder used, has less intensity in the crystalline phases (see brushite and some Wollastonite peaks around 30°). This result can have two contributions: a scale factor that affects XRD for small particles; and a structural reaction (less unreacted Wollastonite) due to the increased Wollastonite surface area, and thus creation of more amorphous calcium phosphate phases.

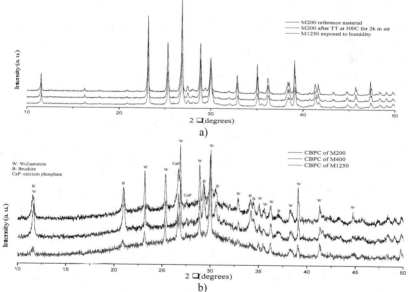

Figure 9 XRD for a) different Wollastonite powders, b) CBPCs made with different Wollastonite powders (as-received).

It is interesting to see how a single phase crystalline material, Wollastonite, after mixing with the phosphoric acid can form a such complex multiphase material with amorphous and crystalline phases in the CBPCs. The amorphous phases are silica and calcium phosphates. They appear in a wide peak with a maximum around $2\theta=30°$. The crystalline phases are Wollastonite (un-reacted from the raw material) and brushite. It is also shown that besides the surface changes (thermal treatments and exposure to humidity) or sizing effects (powder grain size), almost only changes in intensity were detected.

SUMMARY

The explanation of the mechanisms regarding the micro-cracking induced by the thermal drying of water in the CBPCs fabricated in this research has been presented. It was established that the combination of shrinkable and non-shrinkable phases with low interfacial strength in the CBPCs produces shrinkage stresses. Since all materials involved are brittle (Wollastonite, silica, brushite and calcium phosphates), microcracks develop in the matrix (calcium phosphates) and the interfaces.

The effect of particle size distribution of Wollastonite powders on CPBCs has been presented, and it was shown how the curing is changed by exposing the Wollastonite to air moisture, even though its water absorption contents did not show any significant change in both XRD and TGA tests.

From the curing tests it can be observed that manufacturing can become a significant issue when small powders are used, since they react very fast compared with the M200 powder, there is only a very short time to fabricate actual parts (see Figure 5). Wollastonite powders change the curing parameters with the exposure in air. Even though the curing peak position was shifted to a longer time (see Figure 7), no new phases were detected in the XRD and no significant changes were detected at the TGA since weight loss was as low as the instrument limits of detection. We conclude that small changes in the Wollastonite surface can produce high changes in the reactivity when Wollastonite is mixed with the aqueous phosphoric acid formulation. Problems to detect surface changes in Wollastonite have been reported[20].

For the compression tests, it was observed that the error bars increased when the size decreased, while keeping the other manufacturing parameters constant (temperature, mixing time, etc.). On the other hand, compression strength for CBPC made with two year old M200 shows a reduction of about 45% as compared to baseline CBPC fabricated with two week old powder.

Besides the variability and the problem of making samples at room temperature, mean compressive strength results were about 100MPa. However, a higher strength (also lower variability) can be obtained by optimizing the cooling of the raw materials[23].

Density tests show that the lowest values for the CBPCs occur with the finer powders, this behavior was associated with more porous phases as well as with pores introduced during the sample fabrication.

Chemically Bonded Phosphate Ceramics can reach a compressive strength of 100MPa in minutes, whereas Portland cement based concrete reaches a compressive strength about 20Mpa after 28 days. Since the PH is 7.0, reinforcement with (inexpensive) E-glass fibers makes this material superior to traditional cements[22]. Thus, this opens up many applications related to infrastructure repair such as roads, bridges, pipes and pavement repair.

ACKNOWLEDGEMENTS
    The authors wish to thank to the NIST-ATP Program through a grant to Composites and Solutions Inc. (Program Monitor Dr. Felix H. Wu) and to COLCIENCIAS from Colombia for the grant to Henry A. Colorado.

*Corresponding author:
Tel./Fax.: +1-310-206-8157 Email: hcoloradolopera@ucla.edu
Address: MAE 46-127 Engineering IV, 420 Westwood Plaza, Los Angeles, CA 90095, USA

REFERENCES
1. L. C. Chow and E. D. Eanes. Octacalcium phosphate. Monographs in oral science, vol 18. Karger, Switzerland, 2001.
2. Sergey Barinov and Vladimir Komlev. Calcium Phosphate based bioceramics for bone tissue engineering. Trans Tech Publications Ltd, Switzerland, 2008.
3. S. Chattopadhyay. Evaluation of chemically bonded phosphate ceramics for mercury stabilization of a mixed synthetic waste. National Risk Management Research Lab. Cincinnati, Ohio, 2003.
4. D. Singh, S. Y. Jeong, K. Dwyer and T. Abesadze. Ceramicrete: a novel ceramic packaging system for spent-fuel transport and storage. Argonne National Laboratory.Proceedings of Waste Management 2K Conference, Tucson, AZ, 2000.
5. J. F. Young and S. Dimitry. Electrical properties of chemical bonded ceramic insulators. J. Am. Ceram. Soc., 73, 9, 2775-78 (1990).
6. L. Miller and S. Wise. Chemical bonded ceramic tooling for advanced composites. Materials and Manufacturing Processes, Volume 5, Issue 2, 229-252 (1990).
7. A. Gomathi, S. R. C. Vivekchand, A. Govindaraj and C. N. Rao. Chemically bonded ceramic oxide coatings on carbon nanotubes and inorganic nnowires. Adv. Mater., 17, 2757-2761 (2005).
8. M. A. Gulgun, B. R. Johnson and W. M. Kriven. Mat. Res. Soc. Symp. Proc. Vol. 346. Materials Research Society (1994).
9. T. L. Laufenberg, M. Aro, A. Wagh, J. E. Winandy, P. Donahue, S. Weitner and J. Aue. Phosphate-bonded ceramic-wood composites. Ninth International Conference on Inorganic bonded composite materials (2004).
10. A. S. Wagh. Chemical bonded phosphate ceramics. *Elsevier* Argonne National Laboratory, USA. 283 (2004).
11. Simonton Thomas C., Roy Rustum, Komarneni Sridhar, and Breval Else. Microstructure and mechanical properties of synthetic opal: a chemically bonded ceramic . J. Mater. Res. 1, 5, 667-674 (1986).
12. Hu Jiashan, Agrawal D.K. and Roy R. Studies of Strength Mechanism in Newly Developed Chemically Bonded Ceramics in the System $CaO-SiO_2-P_2O_5-H_2O$. Cement and Concrete Research. Vol. 18, 103-108 (1988).
13. Steinke Richard A., Silsbee M. R., Agrawal D. K., Roy R. and Roy D. M. Development of chemically bonded ceramics in the $CaO-SiO_2-P_2O_5-H_2O$ system. Cement and Concrete Research. Vol. 21, 66-72 (1991).
14. S. Y. Jeong and A. S. Wagh. Chemical bonding phosphate ceramics: cementing the gap between ceramics, cements, and polymers. Argonne National Laboratory report, June 2002.

15. Della M. Roy. New Strong Cement Materials: Chemically Bonded Ceramics. February *Science*, Vol. **235** 651 (1987).
16. Wilson, A. D. and Nicholson, J. W. Acid based cements: their biomedical and industrial applications. Cambridge, England, Cambridge University Press (1993).
17. Arun S. Wagh and Seung Y. Jeong. Chemically bonded phosphate ceramics: I, a dissolution model of formation. J. Am. Ceram. Soc., 86, **11**, 1838-44 (2003).
18. Arun S. Wagh and Seung Y. Jeong. Chemically bonded phosphate ceramics: II, warm-temperature process for alumina ceramics. J. Am. Ceram. Soc., 86, **11**, 1845-49 (2003).
19. Arun S. Wagh and Seung Y. Jeong. Chemically bonded phosphate ceramics: III, reduction mechanism and its application to iron phosphate ceramics. J. Am. Soc., 86, **11**, 1850-55 (2003).
20. T.K. Kundu, K. Hanumantha Rao, S.C. Parker. Atomistic simulation of the surface structure of Wollastonite and adsorption phenomena relevant to flotation. Int. J. Miner. Process. **72**, 111–127 (2003).
21. Mosselmans G, Monique Biesemans, Willem R, Wastiels J, Leermakers M, Rahier H, Brughmans S, and Van Mele B 2007 Journal of Thermal Analysis and Calorimetry Vol. **88** 3 723.
22. Colorado H. A, Hahn H. T. and Hiel C. Pultruded glass fiber- and pultruded carbon fiber-reinforced chemically bonded phosphate ceramics. To appear at Journal of Composite Materials. Manuscript ID JCM-10-0398, (2010).
23. Colorado H. A., Hiel C. and Hahn H. T. (2010). Chemically bonded phosphate ceramics composites reinforced with graphite nanoplatelets, Submitted for publication to Composites A. Manuscript ID JCOMA-10-366, (2010).

FOREIGN OBJECT DAMAGE IN AN N720/ALUMINA OXIDE/OXIDE CERAMIC MATRIX COMPOSITE UNDER TENSILE LOADING

D. Calvin. Faucett, Sung R. Choi[†]
Naval Air Systems Command, Patuxent River, MD 20670

ABSTRACT

Foreign object damage (FOD) phenomenon of an N720/alumina oxide/oxide ceramic matrix composite (CMC) was determined with 1.59 mm-diameter hardened steel ball projectiles using impact velocities ranging from 150 to 350 m/s at a normal incidence angle. Target specimens were impacted under tensile loading with three different levels of load factors of 0, 30, and 50 %. Surface damage of impact sites was typically in the form of craters and increased in size with increasing impact velocity. Subsurface damage beneath the impact sites was well developed with fiber/matrix breakage, collapse of pores, delamination, and formation of cone cracks. Difference in FOD between loading and non-loading was significant particularly at 350 m/s with the highest load factor of 50 %, where the targets were on the brink of penetration by the projectiles.

INTRODUCTION

Because of their brittle nature, monolithic ceramics or ceramic matrix composites (CMCs) are susceptible to damage and/or cracking when subjected to impact by foreign objects. This has prompted the propulsion communities to consider foreign object damage (FOD) an important design parameter when those CMCs are intended to be used for structural applications. A significant amount of work on impact damage of brittle materials has been performed experimentally or analytically [1-14], including gas-turbine grade toughened silicon nitrides [15-17].

A span of FOD work has been done ranging from monolithic silicon nitrides [15-17] to various 2D woven CMCs such as melt-infiltrated (MI) SiC/SiC [18], N720™/aluminosilicate (N720/AS) [19], and N720/alumina (N720/A) [20] and 3D woven SiC/SiC [21]. Unlike their monolithic counterparts, 2D or 3D woven SiC/SiC and the oxides/oxides CMCs have not been found to exhibit catastrophic failure for velocities up to 400 m/s, resulting in much increased resistance to impact.

Most of the FOD work in CMCs has been conducted with targets without any external or preloading during impact. The current work describes FOD behavior of an N720/alumina CMC with target samples being tensile-loaded during impact. This is a simulation of a rotating component (e.g. airfoils in aeroengines) in which tensile stresses are generated due to centrifugal force. The oxide/oxide targets in a flat bar configuration were impacted under different tensile loading in a velocity range of 150-350 m/s by 1.59-mm-diameter steel ball projectiles. Damage morphologies were characterized and post-impact strength of impacted targets were assessed. For the authors' best knowledge, this type of FOD work is considered to be unique, applied to CMC material systems.

[†] Corresponding author, sung.choi1@navy.mil

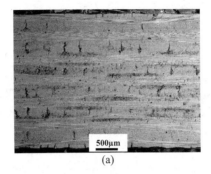

(a)

Figure 1. Microstructure of an N720/alumino oxide/oxide CMC used in this work

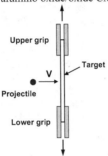

Figure 2. Experimental setup of impact testing under tensile loading for an N720/alumina oxide/oxide CMC

EXPERIMENTAL PROCEDURES

Material and Targets

The CMC used in this work has been described elsewhere [22,20] and is described here only briefly. The composite was a commercial, 2-D woven, N720™ fiber-reinforced alumina matrix oxide/oxide CMC, fabricate by ATK/COIC (San Diego, CA; Vintage 2008). N720™ oxide fibers, produced in tows by 3M Corp. (Minneapolis, MN), were woven into 2-D 8 harness-satin cloth. The cloth was cut into a proper size, slurry-infiltrated with the matrix through a sol-gel process, and 12 ply-stacked followed by consolidation through vacuum-bag technique in low pressure/temperature and pressureless sintering. The fiber volume fraction of the composite panels was about 0.45 as provided by the manufacturer. Typical microstructure of the composite is shown in Fig. 1. No interface fiber coating was applied. Significant porosity and microcracks in the matrix were observed, a characteristic of this class of oxide/oxide CMCs with no fiber interfacial coating primarily for increased damage tolerance [22-24]. Porosity was about 25 %, bulk density was 2.74 g/cm³, elastic modulus was 81GPa determined by the impulse excitation of vibration technique [25]. The flexural or tensile strength of the composite was 140 MPa. Target specimens measuring 12 mm in width, 50 mm in length, and about 2.7 mm in as-furnished thickness were machined from the composite panels.

Foreign Object Damage Testing

Foreign object damage testing was conducted using a ballistic impact gun [15,16]. Briefly, hardened (HRC≥60) chrome steel-balls with a diameter of 1.59 mm were inserted into a 300mm-long gun barrel. A helium-gas cylinder and relief valves were utilized to pressurize and regulate a reservoir to a specific level, depending on prescribed impact velocity. Upon reaching a specific level of pressure, a solenoid valve was instantaneously opened accelerating a steel-ball projectile through the gun barrel to impact a target specimen. The target specimens were tensile loaded through two wedge grips, as shown in Fig. 2. Each target specimen was aligned such that the projectile impacted at the center of the specimen at a normal incidence angle. Three different impact velocities of 150, 250, and 350 m/s were utilized. Applied tensile loading (or stress) during impact was quantified as a load factor (α) defined as

$$\alpha = c_{FOD}/c_f \tag{1}$$

where σ_{FOD} is the applied tensile stress during impact and σ_f is the fracture strength (=140 MPa) of the composite. Three levels of load factor, α = 0, 30, and 50 % were used. For post-impact tests, two or three target specimens were employed at each load factor for a given impact velocity. Three targets were also used for impact (only) with which impact morphologies were to be characterized.

Post-Impact Strength Testing

Post-impact tensile strength was determined for target specimens impacted to better assess the degree of impact damage. Strength testing was carried out via tensile wedge grips with a MTS servohydraulic test frame (Model 312) at a crosshead speed of 0.25 mm/min. As-received tensile strength of the composite was also determined using a total of three dog-boned tensile test specimens.

RESULTS AND DISCUSSION

As seen from the previous study [20], the steel ball projectiles neither exhibited flattening nor visible deformation even at the highest impact velocity of 350 m/s, similar in trend to the case in the N720/aluminosilicate composite [19]. On the contrary, the steel balls impacting harder monolithic silicon nitrides (AS800 and SN282) [15-17] or MI SiC/SiC composite [18] were flattened or severely deformed or fragmented, depending on the level of impact velocity. So, relative to the steel ball projectile, oxide/oxide composites are ascribed as having a 'soft' and porous nature, which may be derived from having a significantly lower elastic moduli of E = 67 and 81GPa for N720/aluminosilicate [19] and N720/alumina respectively, whereas the "hard" and dense silicon nitride and MI SiC/SiC counterparts exhibited values of E = 300 and 220 GPa respectively [15-18].

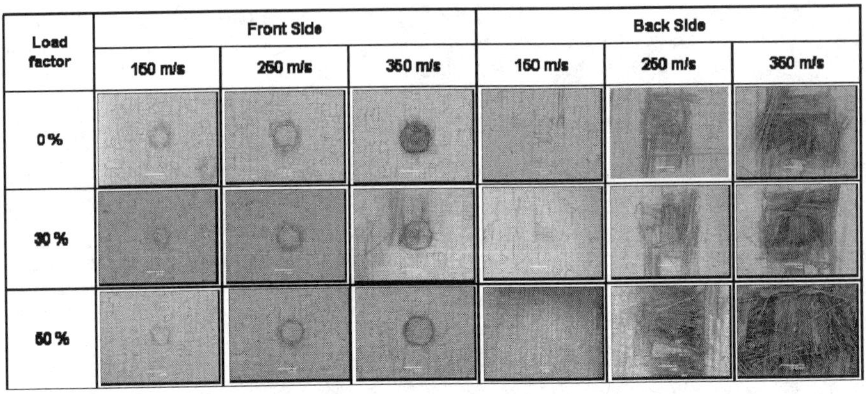

Load factor	Front Side			Back Side		
	150 m/s	250 m/s	350 m/s	150 m/s	250 m/s	350 m/s
0 %						
30 %						
50 %						

Figure 3. Images of frontal and backside impact damages with respect to impact velocity and load factor in N720/alumina oxide/oxide CMC impacted by 1.59-mm steel ball projectiles. Bars=1mm.

Figure 4. Frontal impact damage in a target specimen of N720/alumina oxide/oxide CMC impacted at 350 m/s with a load factor of 50 % by 1.59-mm steel ball projectiles. The direction of tensile loading during impact is indicated as a double arrow. Bars= 200μm.

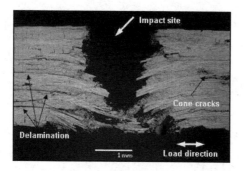

Figure 5. Typical example of the cross-sectional view of a target specimen of N720/alumina oxide/oxide CMC impacted at 340 m/s with a load factor of 30 % by 1.59-mm steel ball projectiles. The direction of tensile loading during impact is indicated as a double arrow.

As observed in the previous work [20], the frontal impact damage of target specimens included craters with fiber/matrix breakage and material removal. The degree of damage or crater size increased with increasing impact velocity as well, as shown in Fig. 3. The effect of load factor on crater size seemed to be insignificant. Figure 4 shows a typical impact site generated at 350 m/s with a load factor of $\alpha = 50$ %. A violent nature of the impact was clearly revealed in the form of severe damage/delamination and breakage of fibers, fiber tows, and matrices.

Figure 3 also shows the backside damage of the target specimens. The backside damage was negligible at 150 m/s regardless of the three load factor conditions tested. For a load factor of 0% intermediate damage was found at 250 m/s progressing to severe damage at 350 m/s, but for impact velocities ≥250 m/s, the degree of backside damage depends on load factor: the greater load factor yields the greater backside damage. It can be also noted that the backside damage was much greater than the frontal damage from V=250 m/s. A typical example of the cross-sectional view of a target specimen impacted at V=350 m/s with α=30 % is shown in Fig. 5. Fiber breakage, backside cracking, delamination, and cone cracking are all seen from the figure. The target, in fact, was on the verge of penetration by the projectile. Notwithstanding, the target specimen as a whole survived without any catastrophic failure.

The results of post-impact strength testing are shown in Fig. 6 in which post-impact tensile strength was plotted with respect to impact velocity as well as to load factor. As-received tensile strength determined, σ_f = 140 MPa, was also included for comparison. It appears that despite some inherent data scatter, post-impact strength decreased with increasing impact velocity for a given load factor. In the similar way, for a given impact velocity the post-impact strength decreased starting from α =30 %. This behavior of post-impact strength degradation is consistent with the results of impact morphologies as already seen from the frontal, backside, and cross-sectional impact damages.

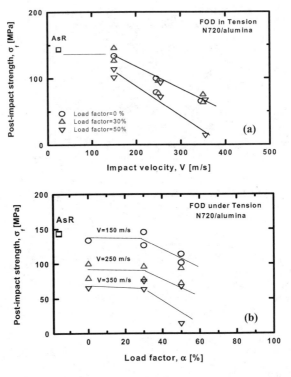

Figure 6. Post-impact strength as a function of impact velocity of N720/alumina oxide/oxide CMC under different levels of tensile loading impacted by 1.59 steel ball projectiles. (a) Post-itrength versus impact velocity; (b) Post-impact strength versus load factor.

The reason for the increased strength degradation with increased tensile loading (α) during impact is due to increased damage or crack size and can be speculated at least qualitatively via a fracture mechanics concept. Impact force is considered as a point load acting on an inherent material flaw and results in damage or crack propagation to the point of equilibrium. When an additional applied stress is engaged during impact, the damage or crack is further increasing until another equilibrium point. The stress intensity factor for the former case (impact force only) may be written in a form [26]

$$K_{I1} = \Phi / c^\kappa \qquad (2)$$

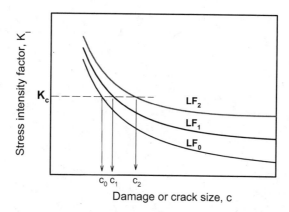

Figure 7. A schematic showing a relationship between stress intensity factor and crack size with different levels of loading. The figure illustrates that with increasing loading factor from $\alpha = LF_0$ to $\alpha = LF_2$, corresponding crack size grows from c_0 to c_2. K_c represents the critical stress intensity factor or fracture toughness of a material.

The stress intensity factor for the latter case (due to tensile loading) may be expressed [26]

$$K_{l2} = \Psi c_{FOD} c^{\lambda} \tag{3}$$

The net stress intensity factor (K_{lt}) is then a combination of Eqs. (2) and (3)

$$K_{lt} = \Phi /c^{\kappa} - \Psi c_{FOD} c^{\lambda} \tag{4}$$

where in Eqs. (2) to (4) σ_{FOD} is applied tensile stress during impact, and c is crack size. The κ, λ, Φ, and Ψ are crack geometry or material dependent parameters. Using Eqs. (1) through (4), a numerical manipulation with arbitrary parameters with $\kappa=3/2$ and $\lambda=1/2$ (typical of a semi-circular or elliptical crack) is shown in Fig. 7, where K_c indicates the critical stress intensity factor or fracture toughness of a material. For a given impact condition, the first term in Eq. (4) remains unchanged. However, the magnitude of the second term increases with increasing load factor from LF_0 (say $\alpha=0$ %) to LF_2 (say $\alpha=30$ %) resulting in increase in crack size from c_0 to c_2 since $K_{lt} \geq K_c$. This increased crack size in turn would degrade further the post-impact strength. With more reliable post-impact strength data and more elaboration in modeling, at least a first-order approximation/prediction of post-impact strength may be feasible using the conventional fracture mechanics concept. However, the results of detailed characterizations of impact damage morphologies should be able to lead to which mechanics -fracture or damage mechanics- is better suited to predict the post-impact strength with respect to the degree of tensile loading.

CONCLUSIONS

The overall impact damage of the N720/alumina oxide/oxide composite was found to be dependent not only on impact velocity but also on load factor. The oxide/oxide composite shows significant impact damage occurring at 350 m/s, particularly for a load factor >30 %. Post-impact strength was consistent with the trend in impact damage in terms of impact velocity and load factor. A schematic to account for post-impact strength (and impact damage as well) as a function of load factor was proposed based on a simplified fracture mechanics concept.

Acknowledgements

The authors acknowledge the support by the Office of Naval Research and Dr. David Shifler.

REFERENCES
1. Wiederhorn, S. M., and Lawn, B.R., 1977, "Strength Degradation of Glass Resulting from Impact with Spheres," J. Am. Ceram. Soc., **60**[9-10], pp. 451-458.
2. Wiederhorn, S. M., and Lawn B. T., 1979, "Strength Degradation of Glass Impact with Sharp Particles: I, Annealed Surfaces," J. Am. Ceram. Soc., **62**[1-2], pp. 66-70.
3. Ritter, J. E., Choi, S. R., Jakus, K, Whalen, P. J., and Rateick, R. G., 1991, "Effect of Microstructure on the Erosion and Impact Damage of Sintered Silicon Nitride," J. Mater. Sci., **26**, pp. 5543-5546.
4. Akimune, Y, Katano, Y, and Matoba, K, 1989, "Spherical-Impact Damage and Strength Degradation in Silicon Nitrides for Automobile Turbocharger Rotors," J. Am. Ceram. Soc., **72**[8], pp. 1422-1428.
5. Knight, C. G., Swain, M. V., and Chaudhri, M. M., 1977, "Impact of Small Steel Spheres on Glass Surfaces," J. Mater. Sci., **12**, pp.1573-1586.
6. Rajendran, A. M., and Kroupa, J. L., 1989, "Impact Design Model for Ceramic Materials," J. Appl. Phys, **66**[8], pp. 3560-3565.
7. Taylor, L. N., Chen, E. P., and Kuszmaul, J. S., 1986 "Microcrack-Induced Damage Accumulation in Brittle Rock under Dynamic Loading," Comp. Meth. Appl. Mech. Eng., **55**, pp. 301-320.
8. Mouginot, R., and Maugis, D., 1985, "Fracture Indentation beneath Flat and Spherical Punches," J. Mater. Sci., **20**, pp. 4354-4376.
9. Evans, A. G., and Wilshaw, T. R., 1977, "Dynamic Solid Particle Damage in Brittle Materials: An Appraisal," J. Mater. Sci., **12**, pp. 97-116.
10. Liaw, B. M., Kobayashi, A. S., and Emery, A. G., 1984, "Theoretical Model of Impact Damage in Structural Ceramics," J. Am. Ceram. Soc., **67**, pp. 544-548.
11. van Roode, M., et al., 2002, "Ceramic Gas Turbine Materials Impact Evaluation," ASME Paper No. GT2002-30505.
12. Richerson, D. W., and Johansen, K. M., 1982, "Ceramic Gas Turbine Engine Demonstration Program," Final Report, DARPA/Navy Contract N00024-76-C-5352, Garrett Report 21-4410.

Functionally Graded Materials

THERMALLY SPRAYED FUNCTIONALLY GRADED MATERIALS

Pavel Chráska, Tomáš Chráska
Institute of Plasma Physics ASCR, v.v.i.
Prague, Czech Republic

ABSTRACT

Single materials are often unable to meet properties required for certain applications. One possible option besides classical composites is functionally graded materials (FGM). Pros and cons of FGM are briefly listed. However, wider application of FGMs depends on availability of suitable production techniques, thermal spraying (TS) being one of them. Both basic types of FGMs, i.e. with a continuous gradient of properties and "sandwich" types can be prepared by TS, such as water stabilized plasma spraying by WSP®. Various FGM produced by plasma spraying are then introduced and their properties discussed. Among them is combination of various oxides, „structural"FGM (having different types of structure at various positions), combination of metals and ceramics, etc. Possible problems in processing as well as of products are pointed out and the target applications proposed.

INTRODUCTION

Further technical development in certain fields is often slowed down or effectively blocked by divergences between visions of designers and the reality of materials engineers. For instance, it is evident that the higher the combustion temperature, the higher the efficiency of the process. However, many materials, such as commonly used steels, "classical" ceramics, etc., are at their properties limits which can be hardly extended. One possible solution to that problem offer composites, made of a matrix and reinforcement with different properties. They have found many interesting and important applications, but in several cases they still do not fully provide solution envisioned by designers. One of the reasons is that ordinary composites, from the "macro" view, exhibit otherwise excellent but rather uniform properties over their volume. Several possible paths exist to overcome this problem, such as various types of coatings, and more recently, the functionally graded materials (FGM). However, new problems are arising at the interfaces of two materials with very different properties, such as ceramic coatings on a steel substrate, or threads in a matrix, etc. Distinct changes in properties at the interfaces can lead to sharp local stress concentrations which are detrimental for the overall performance of a given material/part. These stress concentrations can be eliminated to a high degree if the properties changes are gradual or step-by-step, such as in FGMs.

FGMs are defined as materials with changing microstructure and/or composition across the material's volume. These changes are designed on purpose to cope with different requirements at different parts of the fabricated component[1]. The materials can be designed for specific function and applications.

SHORT OVERVIEW of FGM and their MANUFACTURING

Types of FGM

The architecture of a given FGM is characterized by the course of a certain variable property through the volume. This property can be, for instance, the chemical composition, or phase composition, or microstructure (density of pores, grain size,..), etc. This course can be described by the gradation function, which has two basic types[2]: FGM can be either made of several layers resulting in the step-by-step changes of this function, or a material with continuously changing gradation function can be made (Fig.1).

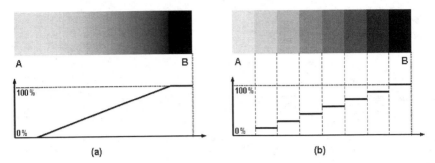

Fig.1. Schematics of FGM types: (a) Continuously changing gradation, (b) multi-layer step-by-step FGM.

Advantages of FGM

Besides fulfilling its main goal, i.e. to supply the required – different – properties at diverse points/surfaces of the future product which, after all, could be achieved by a simple coating, Suresh et al.[1], Bahr et al.[3] and many others (eg. Ref. 4-6) point out various additional advantages of FGM. They list some benefits resulting mainly from the non-existence of a sharp interface in FGMs, such as:

- Decrease of the thermal stresses caused by differences in Coefficients of Thermal Expansion (CTE) of used materials;
- Absence of a sharp interface should lead to an over-all better control of stresses;
- Better cohesion between different materials, such as a metal and a ceramics;
- Extension of the functional region prior to reaching the limiting value of the plastic deformation;
- Suppression of delamination;
- Increase of the fracture toughness;
- Elimination of stress singularities at various locations, such as the free surfaces, corners, roots of cracks, etc.

Disadvantages of FGM

The main obstacle for a wider application of FGMs is their complicated manufacturing due to the mismatch of various properties, e.g. the melting temperature, different affinities to various other elements, particle size, morphology, etc. A typical example is a mixture of a metal with the melting temperature T_{m1} and a ceramics with the melting temperature T_{m2}. In most instances, such a couple is usually $T_{m1} \ll T_{m2}$. Then, if the manufacturing temperature is around T_{m1}, ceramics is still unmelted and the final structure contains unmelted ceramic particles in a metal matrix. In many cases, such structure exhibits a large porosity and a small adhesion between the individual components and its general application is limited. However, for special applications, the presence of very hard particles (carbides) in a tough matrix can be desirable.

On the other hand, working around T_{m2} will definitely result in a rapid oxidation and evaporation of many metals, formation of undesirable phases, etc. Materials must be manufactured in a protective atmosphere or at a low pressure – and both these technologies are quite costly.

Manufacturing of FGM

Various technologies of FGM manufacturing have been offered during the past two decades, mostly based on various modifications of powder metallurgy or thermal spray techniques. A comprehensive general overview up to 2002 is presented in ref. 7, but additional or optimized and re-modified techniques still appear. For instance, there is a patent on FGM processing by cold spraying of ceramic-metal layers, when the final shape is pre-pressed by cold isostatic pressing and it is then sintered using field activated sintering technique (FAST)[8]. Other techniques include, for instance, semi-solid forming process under magnetic field gradients[9], or Centrifugal Mixed-Powder Method[10], etc. However, the aim of this paper is not to give a wide-ranging overview but to concentrate on thermal spray method.

Thermal Spraying of FGM

Thermal spraying (TS) is one of widely used method of manufacturing FGMs coatings. Since ceramics with high T_m is often used as one of the components, utilization of HVOF and similar "low" temperature techniques is rather limited to special cases and plasma based methods are advantageous. During the years there have been published several reports about utilization of various plasma spray techniques, starting with the atmospheric plasma spray (APS), for instance Zhao et al.[11], Tekmen et al.[12], over the low pressure plasma spray (LPPS)[13] to vacuum plasma spray (VPS) – used more for metallic FGMs[14]. According to characterization by Kieback et al.[7], thermal spraying exhibits: very good variability of gradation (or transition) function; versatility in phase content is very good; flexibility in component geometry is also good; both forms of FGM – i.e. coatings and bulk can be prepared, but only thin coatings (10–100 µm) were reported.

However, an additional remark regarding use of TS for FGM production should be made:
It is obvious, that FGM made by TS will inherently contain pores, which will play an important role regarding their over-all properties. Pores are disadvantageous from the point of view of Young's modulus, microhardness, fracture toughness, etc., but they are welcome for their positive effect on the thermal shock resistance and the thermal barrier function. The size distribution and overall volume of pores depends on processing parameters and it can be partly controlled by modification of processing parameters and eventually by an additional treatment like laser glazing, sealing, sintering, etc.

EXPERIMENTAL

Most of the reported samples were prepared using the water stabilized plasma spray (WSP®). Detail description of WSP is given elsewhere[15]. In comparison to classical APS with a gas stabilized gun, WSP offer certain advantages regarding the FGM production. One is the higher temperature and enthalpy of the generated plasma jet permitting to spray materials with high T_m and to reach a larger materials throughput. Other advantage is the possibility of utilizing several external injectors (normally 2 to 4) at the same time and each at its specific own feeding position – see Fig.2. Materials with lower melting point (e.g. Cu and other metals) can be fed in further downstream of the plasma jet while, at the same time, ceramics and other materials with high T_m are injected into the high temperature region of the jet close to the nozzle.

Two types of TS FGM samples were studied: i) Coatings sprayed usually on a plain carbon steel substrates; ii), self-supporting sprayed parts also called free-standing made by a special patented technology[16].

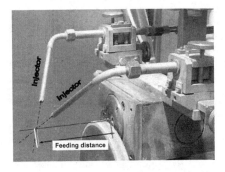

Fig.2. Injector set-up for WSP spraying.

In certain cases, additional treatment was applied, such as controlled heating and cooling of deposits, diffusion annealing or laser after-treatment. The materials used and other experimental details are given in the appropriate place.

Standard techniques for general structural and phase characterization of samples were used, such as SEM with XMA, TEM, XRD, etc. Other particular techniques are introduced whenever needed.

RESULTS and DISCUSSION

The scope of this article does not allow reporting on all of various FGMs systematically produced by WSP. Therefore only selected examples of FGMs are given in the following text. The examples are sorted under five different subheadings.

Sandwiches

Fig.3 shows various types of sandwiches, either made as multilayered deposits or prepared of two distinct materials.

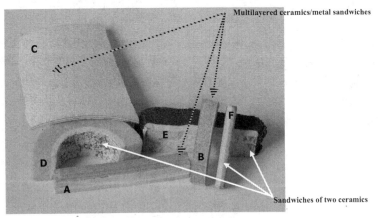

Fig.3. Various FGM free-standing WSP plasma sprayed parts.

Fig.4 presents cross-sections of two examples of the free standing multilayered parts obtained by WSP spraying. A cross-section of a beam (labeled A on Fig.3), made of 10 layers of alumina interlaced with thin metal layers of Aluminum for increased strength, is shown on Fig.4a. Comparison of 3-point bending tests of this multilayered beam and pure aluminabeam of the same thickness and orientation shows that the strength of multilayered beams is up to two times higher.

The other example (C on Figs. 3) shows cross-section (Fig 4b) of a free-standing thin-walled alumina tube interlaced with two layers of Ni-rich alloy. This tube was successfully used as a simple self-contained oven that is heated by induction to reach temperatures of 1100 °C with protective atmosphere inside[17]. It is well known that plasma sprayed deposits are inherently porous. That prevents the use of otherwise advantageous plasma sprayed ceramic tubing wherever gas permeability is unwanted, such as in special furnaces. Metallic coatings are less porous and, in addition, this sandwich type metal layer effectively closes the pores in ceramics.

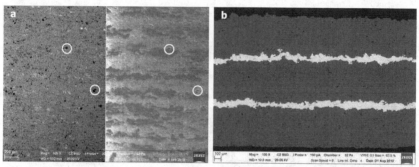

Fig.4. SEM images of cross-sections of multilayered parts from Fig. 3. a) Beam A (backscattered and secondary electrons) images of the same area as denoted by circles around pores; b) Part of tube C (backscattered electrons)

Another example of application of multilayered metal/ceramics FGM is the segment of a transporter from continuous furnaces for heat treatment of special steel parts. The segments were made of AISI 330 steel type. The first plasma sprayed coating was from a Colmonoy type material, the second layer (of approximately double thickness) of mixture of the same metallic powder and zircon (volume ratio 1:1) and the final layer, of approximately 4 fold thickness, from zircon only (the processing details are proprietary). Life time of such treated segments was substantially longer than for uncoated ones[17].

Another type of a sandwich is made of layers of various ceramics (labeled D, E and F on Fig.3). Three different combinations are shown: sample "E" is made of white alumina and a thick coating of garnet almandine. Sample "D" is a section from a tube that contains 1mm thick layer of zircon and several millimeter thick coating of steatite (also called soapstone or talc). The goal in the presented cases was to have materials with varying chemical and/or abrasion resistance on both of the free surfaces[17].

Surface crystallization - "Nano FGM"

This technology is based on the idea of converting amorphous plasma deposits by an aftertreatment[18] into FGM structure. Details of this technology are subject of several patents[18]. The amorphous deposits were prepared mostly of the ternary Al_2O_3-ZrO_2-SiO_2 system containing binary and ternary eutectic and silica as a glass former. A typical composition of the used feedstock was 47.2

wt.% of Al_2O_3, 29.6 wt.% ZrO_2, 19.5 wt.% SiO_2 and 3.7 wt.% of other oxides. This material has been supplied by a Czech company called EUTIT, Ltd., and it is close to their commercial material called "Eucor". WSP was used for plasma spraying of 2 mm thick well-adhered coatings on steel substrates of 25 by 100 mm. The as-sprayed coatings were amorphous.

As reported elsewhere[19] bulk nanocomposites comprising of ZrO_2 nanocrystallites with average crystallite size as low as 12 nm embedded in fine amorphous matrix have been prepared by solid-state crystallization during a controlled heat treatment of as-sprayed parts in a furnace. To produce FGM with specific properties at the surface, the thick amorphous as-sprayed coatings were heat treated on the surface only. Surface heat treatment was performed by pulse Nd:YAG laser JK701H (by Lumonics). Laser was set to energy pulse of 4 J at frequency of 100 Hz and pulse duration of 2 ms. The laser optics was defocused to obtain the laser beam spot of 7,5 mm. Further details of the laser surface heat treatment are also subject to a patent application.

Laser beam was scanned across the sample surface in such a fashion as to achieve temperature above 950 °C at the surface and at the same time prevent rising of temperature above 250 °C at the interface between the coating and the substrate. The high surface temperature sets off crystallization leading to formation of the nanocomposite structure. The low interface temperature is necessary to prevent delamination of the coating due to thermal expansion mismatch. These constraints determined speed of the laser beam as it was scanned across the coating surface. As a result of the laser heat treatment, there is a graded coating with nanocomposite structure at the surface with thickness of approximately 0.6 mm and amorphous structure at the interface with the substrate (Fig.5).

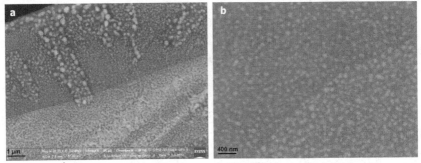

Fig.5. FGM with the nanostructure: (a) Coarser crystallites ~100 nm at the coating surface, and (b) finer nanocrystallites < 100 nm located 200 μm below the surface.

Abrasion resistance of the laser treated graded coating exhibited improvement by 80% in comparison with the as-sprayed coatings. The improvement of abrasion resistance is consistent with the hardness increase and can be attributed to the fine nanocomposite structure of the surface heat-treated samples. Abrasion resistance of the nanocomposite samples is higher than that of plasma-sprayed alumina and comparable or higher than that of chromia sprayed by WSP[20].

Gradual transition FGMs

Coatings with a gradual transition from pure ceramic layer of chromia to pure metallic layer of Ni5Al alloy have also been prepared by WSP. The ceramic and metal feedstock powder were mixed in three different ratios and then fed into WSP torch and sprayed on carbon steel coupons. The resulting coatings are presented in Fig. 6. Measurements of fatigue performance of the coated system revealed that the fatigue life time is extended by more than 50% only if the first plasma sprayed layer is made of ceramics.

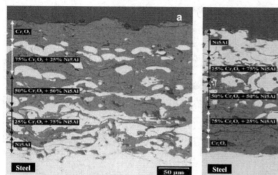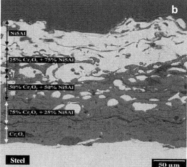

Fig.6. Structure of FGM made of 5 various ratios of Chromia/Ni5Al. (a) First coating layer is metallic, (b) first coating layer is ceramic

Laser glazing

Several papers were published on laser aftertreatment of rather thick plasma sprayed deposits[21-24]. For instance, coatings about 1.5 mm thick from PS Zirconia and gray Alumina were sprayed with WSP and then laser treated with Nd-YAG laser (for details see[21]). Microhardness, surface roughness, reflectivity and slurry abrasion resistance (SAR) were measured before and after the laser treatment. At the surface a distinct layer of 50 – 350 µm thick can be found depending on coatings material and the laser processing parameters. In any combination, however, enhancement of all measured properties was found – for instance, the surface microhardness increased for 40 – 60%.

Deposits of sprayed Corundum with Chromia admixture[22] consisted mostly of softer $\gamma+\delta$ Al_2O_3 phases. When additional post-treatment by a quasi-continuous laser (10-30 W) was applied to the surface, the amount of the hard corundum phase at the surface substantially increased. Careful examination of the surface revealed that the individual splats at and near the surface were fully remelted and the structure is formed as a homogeneous phase with no visible splat boundaries. No transition region between the melted and unmelted zones of the as-sprayed deposits was found – the interface is very sharp and distinct (Fig.7.).

On the other hand, laser treated WSP coatings of Zircon show several transition layers. Zircon on interaction with plasma during spraying decomposed to ZrO_2 and SiO_2. Post-treatment by a quasi-continuous CO_2 laser beam even more altered the as-sprayed deposits. Structure of the as-sprayed samples, containing dissociated amorphous Silica and Zirconia in various forms (tetragonal and monoclinic), has changed and the fine monoclinic dendrites of ZrO_2 were formed. The following layers can be distinguished (see Fig.8):

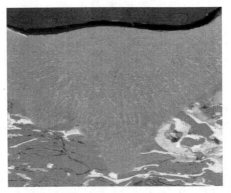

Fig.7. Cross-section of laser treated Al₂O₃-Cr₂O₃ coating.

A.   non-affected structure of as-sprayed splats;
B.   splats with a pronounced very fine columnar structure;
C.   transition layer with a fine structure containing mixture of Zirconia and Silica;
D.   layer with coarse segregated particles of Zirconia and Silica;
E.   monoclinic dendrites of ZrO₂ on the surface.

The main goal of the laser treatment, i.e. obtaining a lower porosity and a lower roughness of the surface, well adhering to the rest of the deposit, has been reached.

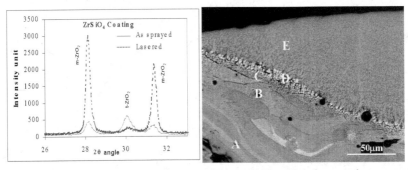

Fig, 8. Laser treated Zircon coating. A) Phase composition   B) Transition phases on the cross-section

Diffusion annealing – Boronizing

Boronizing of metal surface is a technology for increasing the surface hardness and thus the wear resistance, while a secondary goal in some application is also protection against corrosion by molten glass. A new method was proposed combining WSP spraying of a boron-rich material followed by diffusion annealing[25].

Two different steels were used for experiments – plain carbon steel and high-speed tool steel and 3 different boron-rich materials as the powder feedstock for spraying: a) B₄C (78 wt% B); b) Ferro boron (18 wt% B) and c) TiB₂ (30wt% B). The minimal thickness of the plasma sprayed deposit necessary for production of boronized layer of 250 μm has been calculated from stoichiometry as 90-100 μm for B_4C feedstock and more for the others. The diffusion annealing parameters were calculated from the diffusion rates and consequently the temperatures chosen for annealing were 900° to 1000°C for 4 to 8 hours. Metallographic cross-section of the coated steel after diffusion annealing showed various boride phases formed as a consequence of B diffusion into steel. Depending on the primary coatings thickness, none or part of the coating remained at the surface, then phases FeB and Fe_2B subsequently formed, followed by a solid solution of Fe and B and unaltered steel. The most important phase is the typical acicular structure of Fe_2B, providing an ideal duffusional anchoring. These boronized layers can serve as bond coats for remaining unsolved coatings or for secondary plasma sprayed ceramic deposits.

Since the scope of this article does not allow reporting on all of various FGMs produced by WSP only selected examples of FGMs were presented. All of the processes involved in the sample production (including WSP spraying) are fully reproducible. The reproducibility of WSP spraying is given inherently by the used system regardless of the sprayed material.

CONCLUSIONS

Thermal spraying itself or in a combination with an additional treatment is a very efficient and flexible technology of FGM manufacturing. Application of the water stabilized plasma (WSP) then extends the limits or constrains mentioned in the literature for "classical" APS. WSP application allows producing relatively thick deposits – coatings or free-standing bodies – in a reasonable short time using the multiple feeding set-up, inherent to WSP.

Several novel techniques, combining primary WSP spray deposits with additional treatment represent interesting ways of producing FGM i) nanostructured surface layer continuously transitioning into the as-sprayed deposit; ii) boronizing of steel surface with the help of in-diffusion of boron from the B-rich coatings that produces well anchored structure; iii) form of multilayered ceramics/metal sandwiches, exhibiting several interesting properties, such as very limited or no gas permeation trough the part or increased strength.

Utilization of WSP in FGM production could help to broaden present limited application fields to new one, such as various free-standing "tailored" parts for gas management, higher temperature utilization, etc.

REFERENCES

[1] S.Suresh, A.Mortensen, Fundamentals of functionally graded materials, IOM Communications Ltd., London, 1998, 165 p.

[2] U. Schulz, M. Peters, Fr. -W. Bach, G. Tegeder, Graded coatings for thermal, wear and corrosion barriers, *Materials Science and Engineering*, **A362**, 61-80 (2003).

[3] H. -A. Bahr, H. Balke, T. Fett, I. Hofinger, G. Kirchhoff, D. Munz, A. Neubrand, A. S. Semenov, H. -J. Weiss, Y. Y. Yang, Cracks in functionally graded materials, *Materials Science and Engineering*, **A362**, 2-16 (2003).

[4] F.Kroupa, Residual stresses in thick nonhomogenous coatings, *J.Thermal Spray Technology*, **6**, 309-319 (1997).

[5] O.Kesler, J.Matějíček, S.Sampath, S.Suresh, Measurements of residual stress in plasma-sprayed composite coatings with graded and uniform compositions, *Materials Science Forum*, **308-311**, 389-395 (1999).

[6] Y.M.Shabana, N.Noda, K.Tohgo, Combined macroscopic and microscopic thermo-elasto-plastic stresses of functionally graded plate considering fabrication process, *JSME – A: Solid Mech & Mater Eng*, **44**, 362-369 (2001).

[7] B.Kieback, A.Neubrand, H.Riedel, Processing techniques for functionally graded materials, *Materials Science and Engineering*, **A362**, 81–105 (2003).

[8] J.R.Groza, V.Kodash, Methods for production of FGM net shaped body for various applications, United States Patent 7,393,559 issued on Jul 01, 2008.

[9] Tie Liu, Qiang Wang, Ao Gao, Chao Zhang, Chunjiang Wang, Jicheng He, Fabrication of functionally graded materials by a semi-solid forming process under magnetic field gradients, *Scripta Materialia*, **57**, 992-995 (2007).

[10] Y.Watanabe, Y.Inaguma, H.Sato, E.Miura-Fujiwara, A Novel Fabrication Method for Functionally Graded Materials under Centrifugal Force, *Materials*, **2**, 2510-2525 (2009).

[11] H. X. Zhao, T. Masuda, C. L. Li, T. Takahashi, and M. Matsumura, Corrosion and Erosion-Corrosion of Ceramics and Functionally Gradient Material-Coated Steels in Erosion Environments, *Corrosion* **56**, 654-665 (2000).

[12] C.Tekmen, I.Ozdemir, E.Celik, Failure behaviour of functionally gradient materials under thermal cycling conditions, *Surf. Coat. Technol.*, **174–175**, 1101–1105 (2003).

[13] S.Jiansirisomboon, K.J.D.MacKenzie, S.G.Roberts, P.S.Grant, Low pressure plasma-sprayed Al2O3 and Al2O3/SiC nanocomposite coatings from different feedstock powders, *J. Europ. Cer. Soc.* **23**, 961-976 (2003).

[14] J.-E. Döring, R.Vaßen, G. Pintsuk, D. Stöver, The processing of vacuum plasma-sprayed tungsten–copper composite coatings for high heat flux components, *Fusion Engineering and Design*, **66-68**, 259-263 (2003).

[15] P.Chráska, M.Hrabovský, Proc.Inter.Thermal Spray Conf. ITSC'92, Orlando, USA, 1992, Ed.: C.C.Berndt, ASM Inter., Materials Park, USA, p. 81 - 85.

[16] K.Neufuss, P.Chráska, Czech Patents No. PV 2038-96 and 2883-96, 1996.

[17] K.Neufuss, IPP ASCR, Prague, unpublished proprietary results 1995-2000.

[18] T.Chráska, K. Neufuss: Czech Patent No CZ 300602 B6 (2009) and PV 2009-307 / 18.5.2009;

[19] T. Chraska, K. Neufuss, J. Dubsky, P. Ctibor, and M. Klementova, Fabrication of Bulk Nanocrystalline Ceramic Materials, *J.Thermal Spray Technology*, **17**, 872-877 (2008)

[20] J..Nohava and P. Chráska, Wear Resistance of Al2O3 and Cr2O3 Coatings Deposited by Water-Stabilized Plasma Spraying. In: Proc.International Symposium on Plasma Chemistry, R. d'Agostino, Ed., Int. Plasma Chemistry Society, 2003. Proc. 16th Inter. Symp. Plasma Chemistry, Taormina, Italy, 2003, Ed.: d'Agostino R.; Favia P.; Frascassi F.; Palumbo F. IUPAC 2003, p.651-6, CD Rom

[21] P.Ctibor, J. Nohava, I.Karovicova, P.Chráska, J.Tuominen, P.Vuoristo, Improvement of mechanical properties of Alumina and Zirconia plasma sprayed coatings induced by laser after-treatment, Proc.Inter. Ther. Spray Conf. ITSC 05, Basel 2005, Switzerland, DSV-Verlag Dusseldorf, Ed. E.Lugscheider, ISBN 3-87155-793-5, p.1033-1038.

[22] J.Dubský, B.Kolman, P.Chráska, A.Jančárek, Laser post-treatment of plasma sprayed Al2O3-Cr2O3 coatings, Proc. Inter.Thermal Spray Conf. ITSC 06, Seattle, WA, USA, 2006, Ed.:B.R.Marple, C.Moreau, ASM Inter., Materials Park, USA, CD-ROM 5_7_11989.pdf.

[23] P.Ctibor, L.Kraus, J.Tuominen, P.Vuoristo, P.Chráska, Improvement of mechanical properties of Alumina and Zirconia plasma sprayed coatings induced by laser post-treatment Ceramics - Silikáty **51**,181-189 (2007), ISSN 0862-5468.

[24] B.Kolman, K.Neufuss, A.Jančárek, J.Dubský, P.Chráska, Plasma sprayed zircon deposits modified by laser treatment, Proc. Conf. on Dense Plasma Interaction with Materials, Ed. U. Ugaste, Uni Tallin, 2003, p. 69-78.

[25] V. Brožek, V. Dufek, P.Chráska, Plasma sprayed gradient materials with boride interlayer, Proc. Euro Conf. on Advances in Hard Materials Production, Stockholm, Sweden, 1996, p. 451 – 453.

ACKNOWLEDGEMENT
Partial support of the VZ0 Z20430508 and of the Czech Science Foundation under grant No. GACR 106/08/1240 is gratefully acknowledged. The authors would like to express their thanks to their present or former colleagues, Messrs. K.Neufuss, J.Bensch, V,Brožek, B.Kolman, J.Dubský, P.Ctibor, J.Matějíček, R.Mušálek and others for their help.

Ceramic Processing

DEVELOPMENT OF ULTRA-HIGH TEMPERATURE STABLE CERAMICS BY REACTIVE INFILTRATION PROCESSES

R.Voigt, W.Krenkel, G.Motz
Ceramic Materials Engineering, University of Bayreuth
Bayreuth, Germany

A. Can
Element Six$_{TM}$
Springs, South Africa

ABSTRACT

In recent years there has been an increasing demand for materials that can operate at temperatures higher than 2000 °C. Ultra-high temperature ceramics like refractory carbides fulfill this requirement. Among these materials Hafniumcarbide (HfC) is of special interest due to its extremely high melting point and good chemical stability at high temperatures. Refractory carbides are typically produced by hot pressing at high temperatures and high pressures.

A new approach is the formation of HfC via a reactive melt infiltration process. Because of the high melting point of pure Hafnium (2233°C) Hf-alloys like HfSi8at%, HfV43at%, TiHf20at% and HfMo34at% with a melting point lower than 1850°C are used in this work. A pre-condition for the use of these alloys in a melt infiltration process is their reactivity with carbon to form carbides.

For reactivity tests tablets of carbon powder, a carbon precursor (phenolic resin) as binder and alloy granules were pressed, cured, pyrolysed at 1000°C under Ar-atmosphere and subsequently annealed over the melting point of the respective alloy . The characterization of the annealed samples with XRD revealed the formation of HfC for all alloys used except for the TiHf-alloy. Additionally the carbides of the respective alloying element are formed: SiC, V_xC_y, TiC and Mo_2C. First results show that there is an influence of the annealing parameters (temperature, time) and the carbon modification (amorphous carbon, graphite) on the microstructure and the crystalline phases formed for the system HfV.

INTRODUCTION

Since the late 1960s materials for high temperature applications are limited to Siliconcarbide and Siliconnitride[1]. Recently, Ultra-High-Temperature-Ceramics (UHTC) have gained significant interest for the use in aerospace as well as energy applications at temperatures higher than 2000 °C[2]. The term UHTC is used for refractory compounds such as ceramic borides, nitrides and carbides which are characterized by a high melting point, chemical inertness and a good oxidation resistance at extreme application conditions. Within the group of refractory carbides, Hafniumcarbide (HfC) is of special interest due to its extremely high melting point (3928°C) as well as its high hardness and excellent oxidation stability at temperatures higher than 1800 °C [5-7].

Despite their potential, Ultra-High Temperature Carbides have not been developed on an industrial scale because of their low sinterability. The processing is based on hot- or hot-isostatic pressing[3]. Opeka et al. report processing conditions of 2540 °C and 16.5 MPa for $HfC_{0.67}$[8]. It is obvious that these processing conditions limit the size and shape of UHTC parts and increase their costs significantly[2]. Therefore, the development of a pressureless process that operates at lower temperatures is of great importance for the opening of new fields of applications for Ultra-High Temperature Ceramic Carbides.

A new approach is the development of Ultra-High Temperature Carbides via reactive melt infiltration. This process is already used in industry for the production of Silicon Carbide or carbon fibre reinforced Silicon Carbide. During the reactive melt infiltration liquid silicon and solid carbon are converted to Silicon Carbide[9]. This process is characterized by lower processing temperatures in comparison to the conventional hot pressing and nearly no dimensional changes during processing[9,10]. A basic requirement for the application of this technology is the reactivity of the used metal or the used alloy with carbon.

In this work we report about first results to investigate the reactivity of four different hafnium alloys with carbon. For this purpose tablets of four different hafnium alloys with a melting point between 1338°C and 1820 °C, carbon powder and a carbon precursor (phenolic resin) as binder are pressed, pyrolysed and subsequently annealed over the melting point of the respective alloy. The following characterization was carried out regarding the generated crystalline phases and the microstructure formed. The influence of the annealing time, annealing temperature and carbon modification was investigated.

EXPERIMENTAL PROCEDURE

Table I shows the composition of the hafnium alloys used in this work (Hauner Metallische Werkstoffe, Germany), each with a grain size smaller than 1 mm. As reaction partner commercial available amorphous carbon powder (PC40, Timcal Ltd., Switzerland) with an average grain size of $d_{50}=15.6\mu m$ or graphite powder (GSI 45, Remacon, Germany) with an average grain size of $d_{50} = 5.96\mu m$ was chosen. A carbon precursor (phenolic resin 6227 FP, Bakelit AG, Germany) was added to act as binder for the processability of the samples and as an additional carbon source for the carbide formation.

Table I. Composition of the Hafnium alloys used in this work

Hf - percentage	SiHf	HfV	TiHf	HfMo
atomic-%	8	43	20	34
weight-%	36	82	45	78

The pressing mixtures consist made of the respective alloy granules and 1.5 times the amount of carbon needed for a stoichiometric conversion to enable a quantitative carbide formation. The binder fraction was 30vol% for every mixture. After the cold isostatic pressing of the samples at 150MPa, the samples were cured at 120 °C under argon atmosphere and subsequently pyrolysed at 1000°C for 30min under argon atmosphere (GERO GmbH, Germany). The annealing of the samples for the carbide formation was carried out in argon atmosphere in a graphitic heated box furnace (HTK 8, Gero GmbH, Germany). Temperatures for the annealing were chosen to be over the melting point of the respective alloy to support the carbide formation. To get a raising of the melting point which is relatively seen the same for each alloy, the maximum annealing temperature was set as 1/6 and 1/4 over the melting point of the respective alloy. Table II shows the melting temperatures of each alloy and the resulting annealing temperatures. The annealing time was varied between 1, 2.5 and 5 hours.

Table II. Melting point and annealing temperature of each Hafnium alloy

	SiHf8at%	HfV43at%	TiHf20at%	HfMo34at%
Melting point T_m [°C]	1338	1456	1650	1866
Annealing Temperature T_{A1} (T_m+1/6T_m) [°C]	1561	1699	1925	2177
Annealing Temperature T_{A2} (T_m+1/4T_m) [°C]	1673	1820	2063	- - -

The samples were characterized regarding their microstructure after pyrolysis and after annealing with optical and scanning electron microscope (SEM Jeol, JSM 6400). For elemental analysis an energy dispersive X-ray spectroscopy system (EDS, Soran System Six Model 300, Thermo Fisher) was used. The crystalline phases formed were determined with X-ray diffraction (XRD, Philips X'Pert, PANanalytical).

RESULTS AND DISCUSSION

Reactivity and Microstructure of the system SiHf8at%
    Both alloying elements in the system SiHf8at% have a large negative enthalpy of formation for carbides (Table III).

Table III. Enthalpy of formation for carbides in the system Si-Hf[4]

	HfC	SiC
Enthalpy of formation [kJ/mol]	- 209.6	-68

The XRD measurements showed the formation of HfC and SiC for each experiment (Table IV). No influence of the annealing temperature, annealing time or carbon modification can be seen in the XRD results. Since no starting phases (Si and $HfSi_2$) are detectable it can be assumed that the conversion of the starting phases is already completed at the lowest annealing temperature (1561°C) and the shortest annealing time (1h) independent of the carbon modification used as reaction partner.

Table IV. Formation of crystalline phases in the system SiHf8at% in dependence of annealing temperature, time and reaction partner

Hf alloy	T_{max} [°C]	Carbon	Annealing time [h]		
			1	2.5	5
SiHf8at%	1561	amorphous	HfC, SiC, C	HfC, SiC, C	HfC, SiC, C
SiHf8at%	1561	graphitic	HfC, SiC, C	HfC, SiC, C	HfC, SiC, C
SiHf8at%	1699	amorphous	HfC, SiC, C	HfC, SiC, C	HfC, SiC, C
SiHf8at%	1699	graphitic	HfC, SiC, C	HfC, SiC, C	HfC, SiC, C

Figure 1a shows the microstructure after pyrolysis at 1000 °C. As expected, the alloy granules are located separately in a carbon network which consists of the pyrolysed phenolic resin and the carbon powder added. No melting occurred during pyrolysis because the temperature is below the melting temperature of the alloy. In contrast, Figure 1b shows a microstructure after annealing of the sample at 1561°C (2.5h). It can no longer be distinguished between separate alloy particles, the particles melted and spread. The liquid alloy infiltrated into the carbon matrix and reacted with the carbon to form carbides. HfC, SiC and C, the three phases detected with XRD can clearly be seen in Fig. 1c. It is noticeable that HfC is located at the interphase to carbon. According to the enthalpies of formation (Table III), Hf should react first with the carbon to form the HfC ceramic. Further diffusion processes through the HfC at the interface should lead to a complete conversion of the alloy into HfC and SiC even in the core.

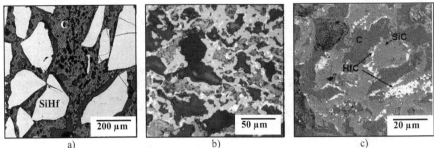

a)   b)   c)

Figure1. System SiHf8at% a) after pyrolysis (1000°C, 30min, Ar), b) after annealing (1561°C, 2.5h, Ar) and c) EDS phase analysis

Reactivity and Microstructure of the system HfV43at%

The results of the XRD measurements show an influence of the annealing temperature and time as well as of the carbon modification on the crystalline phase formation in the system HfV43at% (Table V). Starting material (HfV$_2$) was still found after annealing at 1699°C and 1h respectively 2.5h holding time for graphitic carbon as reaction partner. To get a complete conversion at a temperature of 1699°C an annealing time of 2.5 hours is necessary with amorphous carbon and 5h with graphitic carbon as reaction partner. At 1820°C even 1h holding time leads to a complete conversion of the starting phase HfV$_2$, because it is no longer detectable with XRD measurements. For both temperatures a dependence of the carbide stoichiometry on the annealing time and on the carbide modification can be observed. With amorphous carbon as reaction partner V$_8$C$_7$ is formed first but after 5h holding time VC is the stable carbide for both annealing temperatures. With graphitic carbon as reaction partner the carbide formation starts with V$_4$C$_3$ for both annealing temperatures but the stable modification after 5h annealing is V$_8$C$_7$.

Table V. Formation of crystalline phases in the system SiHf8at% in dependence of annealing temperature, time and reaction partner

Hf alloy	T$_{max}$ [°C]	Carbon	Annealing time [h]		
			1	2.5	5
HfV43at%	1699	amorphous	HfV$_2$, HfC, V$_8$C$_7$,C	HfC, V$_2$C, V$_8$C$_7$,C	HfC, VC, C
HfV43at%	1699	graphitic	HfV$_2$, HfC, V$_4$C$_3$, C	HfV$_2$, HfC, V$_4$C$_3$, C	HfC, V$_8$C$_7$; C
HfV43at%	1820	amorphous	HfC, V$_8$C$_7$; C	HfC, V$_8$C$_7$; VC, C	HfC, VC, C
HfV43at%	1820	graphitic	HfC, V$_4$C$_3$, C	HfC, V$_4$C$_3$, C	HfC, V$_8$C$_7$; C

The microstructure after pyrolysis at 1000 °C is characterized by separated alloy particles embedded in a carbon network (Fig. 2a). The microstructure after annealing at 1699 °C is dependent on the annealing time. For 1h holding time the alloy particles are still separate (Fig. 2b) but they have clearly changed their shape due to melting. One possible explanation for the non-spreading behavior is the formation of a passivating carbide layer around the particles, so that longer annealing times or higher annealing temperatures are necessary for the carbide formation. After annealing for 5h the particles melted and infiltrated the carbon matrix (Fig. 2c). The microstructure of the samples annealed at 1820°C is comparable to the microstructure after annealing at 1699°C and 5h annealing time (Fig. 2c). For all different annealing experiments two different carbide phases were observed. HfC is again located at the interface to the carbon matrix, whereas V$_x$C$_y$ can only be found in direct contact to the

HfC but not to carbon (Fig. 2b). Based on the enthalpies of formation the formation of HfC (-209.6kJ/mol[4]) is favored over the formation of VC (-100.8 kJ/mol[4]). The formation of V_xC_y in the core suggests diffusion processes inside the particles.

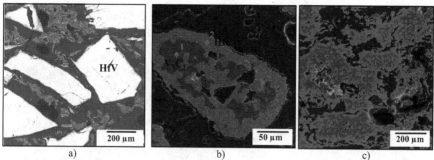

a)   b)   c)

Figure 2. System HfV43at% a) after pyrolysis (1000°C, 30min, Ar), b) after annealing at 1699°C for 1h in Ar and c) after annealing at 1699°C for 5h in Ar.

Reactivity and Microstructure of the system TiHf20at%

Whereas after pyrolysis the TiHf20at% alloy particles are located separately in a carbon matrix (Fig. 3a), microstructural characterization after annealing showed the formation of a network for every annealing experiment (Fig 3b). With regard to the similar enthalpies of formation a simultaneous formation of HfC (-209.6 KJ/mol)[4] and TiC (184.3 kJ/mol)[4] was expected. But Figure 3c with higher magnification in combination with elemental analysis (EDS) shows a separation of the alloy into Ti and Hf during annealing. Furthermore a thin TiC layer was formed at the interface between Titanium and Carbon whereas the Hafnium is located in the core without direct contact to the carbon of the sample. One possible reason for the observed phase formation is e.g. the different wetting behavior of Ti and Hf with carbon. Krenkel et al.[11] reported a similar behavior for the infiltration of Cu-Ti alloys into carbon fiber reinforced carbon. They also observed a separation of the alloy into Ti and Cu and a formation of TiC at the interface.

a)   b)   c)

Figure 3. System TiHf20at% a) after pyrolysis (1000°C, 30min, Ar), b) after annealing at 2063°C for 2.5hh in Ar and c) EDS analysis

Table VI shows the detected crystalline phases after annealing of the TiHf20at% samples. No crystalline HfC was detected even for 5h annealing at 2063°C. This allows the conclusion that no diffusion processes occur to form crystalline HfC.

Table VI. Formation of crystalline phases in the system TiHf20at% in dependence of annealing temperature, time and reaction partner

Hf alloy	T $_{max}$ [°C]	Carbon	Annealing time [h]		
			1	2.5	5
TiHf20at%	1925	amorphous	Hf, TiC;C	Hf, TiC, C	TiC, C
TiHf20at%	1925	graphitic	Hf, TiC, C	Hf, TiC, C	TiC, C
TiHf20at%	2063	amorphous	n.a.*	TiC, C	TiC, C
TiHf20at%	2063	graphitic	n.a.*	TiC, C	TiC, C

Reactivity and Microstructure of the system HfMo34at%
After pyrolysis at 1000°C for 30min under argon the microstructure for the System HfMo34at% is comparable to the other investigated systems. The alloy particles are embedded in a carbon matrix (Fig. 4a). After annealing at 2177°C the metal phase spreads over the carbon network (Fig. 4b). A more detailed look at the brighter metal phase in Figure 4b shows the segregation into two phases. With the help of EDS measurements they can be identified as Mo-rich (dark grey) and Hf-rich (light-grey) regions (Fig. 4c). No carbide formation was observed in the inner part of the metal containing regions which suggests the conclusion that the carbide formation only took place at the interface to the carbon network.

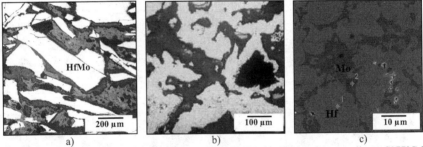

a)   b)   c)

Figure 4: System HfMo34at% a) after pyrolysis (1000°C, 30min, Ar), b) after annealing at 2177°C for 2.5h in Ar and c) EDS analysis of the bright region in fig. 2b

X-ray diffraction analysis shows the presence of HfC and Mo$_2$C for each sample after annealing at 2177 °C (Table VII). But also unreacted Hf was detected so the conversion into a carbide was not completed. This result is not in accordance with the enthalpies of formation, because the formation enthalpy of Mo$_2$C (-46 kJ/mol)[4] is much lower than the value for HfC (-209.6 KJ/mol)[4] so complete conversion of the Hf was expected. The carbon source has no influence on the phase formation.

---

* samples were not processed

Table VII. Formation of crystalline phases in the system HfMo34at% in dependence of annealing temperature, time and reaction partner

Hf alloy	T $_{max}$ [°C]	Carbon	Annealing time [h]	
			1	2.5
HfMo34at%	2177	amorphous	HfC, Mo$_2$C, Hf, C	HfC, Mo$_2$C, Hf, C
HfMo34at%	2177	graphitic	HfC, Mo$_2$C, Hf, C	HfC, Mo$_2$C, Hf, C

CONCLUSIONS

The present work shows that low melting Hf-alloys SiHf8at%, HfV43at% and MoHf34at% are suitable for the formation of HfC by a reaction between the molten alloy and carbon. Additionally carbides of the respective alloying elements are formed: SiC, V$_x$C$_y$ and Mo$_2$C.

The system SiHf8at% is the most reactive. The reaction to HfC and SiC is already completed after annealing at 1561 °C (1/6 over the melting point of the alloy) and 1h holding time. In contrast to that, the carbide formation in the System HfV43at% depends on the annealing temperature, time and the carbon modification used as reaction partner. For amorphous carbon VC is the stable phase after annealing for 5h and for graphitic carbon V$_8$C$_7$ is the stable phase after 5h annealing. At the lower annealing temperature (1699 °C) an annealing time longer than 1h is necessary for a complete conversion.

The microstructure after annealing is similar for all systems, but the formation of the carbides depends on the alloy used. The alloy particles, which were separate after pyrolysis, melted together and spread over the sample. Only for the HfV43at% system annealed at 1699°C for 1h the particles did not spread over the sample. However the change in shape after annealing allows the conclusion that melting occurred. For the system SiHf8at% and HfV43at% a preferential location of HfC at the interface to carbon can be observed. For both systems the enthalpy of formation for HfC is higher than for SiC and VC respectively, so Hf reacts first. The fact that a formation of SiC and V$_x$C$_y$ nevertheless can be observed reveals diffusion processes through the HfC layer. For the system TiHf20at% a complete segregation into a Ti- and a Hf-phase with the formation of a thin TiC layer at the interface to carbon was found .However the enthalpy of formation is slightly higher for HfC than for TiC. So other effects like e.g. better wetting must play a role in this system.

REFERENCES
[1] E. Wuchina, E. Opila, M. Opeka, W. Fahrenholtz and I. Talmy, UHTCs: Ultra-High Temperature Ceramic Materials for Extreme Environment Applications, *The Electrochemical Society Interface*, **2007**, 4, 30-36
[2] Y.D. Blum, S. Young, D. Hui and E. Alvarez, Hafnium Reactivity below 1500°C in search of better processing of HfB$_2$/SiC UHTC composites, in *Ceramic Engineering and Science Proceedings*, 27, 2, *30th International Conference on Advanced Ceramics and Composites*, **2006**, p 617-628
[3] M.J. Gasch, D.T. Ellerby and S.M. Johnson, Ultra High Temperature Ceramic Composites, in *Handbook of Ceramic Composites* (Ed.: N.P. Bansal), Kluver Academic Publishers, New York, NY, USA, **2005**, Chapter 9
[4] H.O. Pierson, Handbook of refractory carbides and nitrides, Noyes Publications, Westwood, New Jersey, USA, **1996**
[5] D. Sciti, L. Silvestroni, A. Bellosi, High-Density Pressureless-Sintered HfC-Based Composites, *Journal of the American Ceramic Society*, **2006**, 89(8), 2668-2670
[6] A. Sayir, Carbon fiber reinforced hafnium carbide composite, *Journal of Materials Science*, **2004**, 39, 5995-6003

[7] E. Wuchina, M. Opeka, S. Causey, K. Buesking, J. Spain, A. Cull, J. Routbort, F. Guitierrez-Mora, Designing for ultrahigh-temperature applications: The mechanical and thermal properties of HfB_2, HfC_x, HfN_x and $\alpha Hf(N)$, *Journal of Materials Science,* **1999**, 39, 5939-49.

[8] M.M. Opeka, I.G. Talmy, E.J. Wuchina, J. A. Zaykoski, S.J. Causey, Mechanical, Thermal, and Oxidation Properties of Refractory Hafnium and Zirconium Compounds, *Journal of the European Ceramic Society,* **1999**, 19, 2405-2415

[9] W. Krenkel, Carbon Fibre Reinforced Silicon Carbide Composites (C/SiC, C/C-SiC), in *Handbook of Ceramic Composites* (Ed.: N.P. Bansal), Kluver Academic Publishers, New York, NY, USA, **2005,** Chapter 5

[10] M.Singh, D.R. Behrendt, Reactive melt infiltration of silicon-niobium alloys in microporous carbon, *J. Mater. Res.* **1994**, Vol 9, No.7, 1701-1708

[11] J. Schmidt, M. Fries und W. Krenkel, Herstellung eines Verbundwerkstoffs durch Schmelzphaseninfiltration von binären Kupfer-Titan Legierungen in kohlenstofffaserverstärkten Kohlenstoffstrukturen, in *Verbundwerkstoffe und Werkstoffverbunde 2001* (Ed. B. Wielage), Wiley VCH , Weinheim, **2001**

ACKNOWLEDGEMENTS
    The German Science Foundation (Deutsche Forschungsgemeinschaft, DFG) is acknowledged for financial support (Research Training Group 1229) and Element Six_{TM} Springs, South Africa where some of this work has been done.

SHAPE EVOLUTION OF SPANNING STRUCTURES FABRICATED BY DIRECT-WRITE ASSEMBLY OF CONCENTRATED COLLOIDAL GELS

Cheng Zhu and James E. Smay
School of Chemical Engineering, Oklahoma State University, Stillwater, Oklahoma, 74078

ABSTRACT

This paper proposed two models, a purely empirical one and a mathematical one, which described the geometric fidelity of 3D mesoscale lattice structures through direct-write assembly of colloidal gels. Although colloidal gels have excellent self-support properties, their transient viscoelastic dynamics inevitably leads to fast deflection of spanning elements under gravity. The empirical model is based on the dimensional analysis of related process variables. Dimensionless groups were obtained through Buckingham Pi theorem, and organized in a non-monomial form to predict the maximum deflection of span midpoint. The second mathematical model was derived from static force analysis of spanning elements by assuming mechanical behavior of colloidal gels to follow a Kelvin-Voigt analogue. This model resembled viscoelastic catenary to track the deflection evolution of spanning elements after deposition, and computes their equilibrium shape profiles after long resting time. Both of these two models were tested against published experimental data of PZT colloidal inks, which confirmed the validity of these two models in a universal context.

INTRODUCTION

Direct-write assembly [1, 2] of colloidal gels [3-7] enables the fabrication of 3D mesoscale (10^{-2}~10^{-4} m) periodic structures [3, 4] by extruding ink filaments in a layer-by-layer scheme. So far, a number of patterns, such as space-filling layers, high aspect ratio walls, and especially lattice structures (Figure 1) [5] have been manufactured and found with potentially applications in photonic band gap structures [6], artificial bone structures [6, 9, 10], and metal-ceramic composites [6-8]. The geometric fidelity of these structures is very important to their physical function and this fidelity has been closely tied to inks rheological properties [3]. Although colloidal inks have demonstrated excellent self-supporting features for 3D periodic structures assembly [9], a thoroughly fundamental study is needed to discover interrelationships between colloidal processing, ink properties and deposition history.

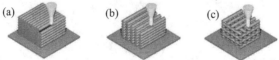

Figure 1 Schematic illustration of 3D periodic structures: (a) space filling layers, (b) high aspect ratio walls, and (c) lattice structures.

The deflection of span filaments has been modeled by a number of investigators. Smay et al., [5] first related equilibrium shear rheological properties of PZT inks to the shape of circular spanning filaments by applying static Euler-Bernoulli beam theory with an effective elastic modulus G'_{eff}. The minimum G'_{eff} required having a mid-span deflection of less than 5% of the filament diameter is given by [5]:

$$G'_{eff} > 1.4\rho_{gel}gs^4D \qquad (1)$$

where ρ_{gel} is the gel density, g is the acceleration of gravity, s = L/D is the span distance (L) divided by the filament diameter (D). This model was then applied to analyze the deflection of spanning filaments fabricated by ceramic inks [10, 11], organic inks [12], and fugitive inks [13]. The relation of equilibrium shear properties to final shape is a simple view of the actual processes happening during the extrusion steps. During the extrusion, the ink must sustain the creep deformation caused by the extrusion pressure induced shear stress. The spanning filaments inevitably undergoes sagging deformation due to capillary and gravity induced tensile stresses after deposition prior to the final solidification step [5]. Therriault et al., [14] modified this quasi-static Euler-Bernoulli equation to a dynamic form by replacing the elastic modulus with the time-dependent tensile creep compliance. However, the model can only provide good predictive values for high aspect ratio spanning filaments with length-to-diameter L/D > 20. There are two limitations, which make the beam theory inaccurate for predicting the deflection of spanning filaments from colloidal inks. First, the mesoscale spanning filaments fabricated by direct-write assembly generally exhibit low aspect ratio (i.e., L/D < 10), which is beyond the beam theory assumptions (i.e., L/D > 20). Second, colloidal inks belong to viscoelastic materials, whose viscous properties were neglected in beam theory. During the deposition of spans, the extruding filament bends 90 degree upon exiting the nozzle to traverse the gap between to supports. That means the elements have both viscous flexibility and elastic flexural rigidity.

Previously, Smay et al., [5] measured the equilibrium shape of spanning elements with different span distances by noncontact laser profilometry scan. In this paper, we first use dimensional analysis to reduce several process variables into three dimensionless groups. An empirical equation was used to relate these dimensionless groups. By comparing with experimental data, this model was verified to be able to predict the maximum deflection of mid-span point. Besides, a time-dependent viscoelastic catenary model was developed to simulate the dynamic shape evolution process. The dynamic evolution process is correlated to the viscoelastic characteristics of colloidal inks and operation parameters. This model has explicit physical meaning, and more precise prediction ability. The whole span filament deflection profiles can be predicted, and the shape evolution with time can also be predicted.

EMPIRICAL MODELING USING DIMENSIONAL ANALYSIS

Dimensionless Groups

Although previous simply supported beam model [5] is an idealized view of spanning phenomena, it is still possible to provide useful information about impact factors on deformation of spanning elements. When using this model to fit deflection profiles of spanning elements, the maximum deflection of mid-span δz_{mid} can be written as:

$$\delta z_{mid} = -\frac{5wL^4}{384EI} \qquad (2)$$

where w is the distributed load (= $0.25\rho_{gel}g\pi D^2$), the Young's modulus E is determined by the shear modulus G' with the relationship E = $2(1+v)G'$ [15], where v = 0.5 is the Poisson coefficient [16], and I is the area moment of inertia of the circular cross section (= $\pi D^4/64$). In this model, the suitable G'_{eff} was selected to replace the G'. Although the relationship between G'_{eff} and other variables has been demonstrated, it is still not a direct method to disclose how the different factors influence the deflection degree of the spanning elements. Here, we assume that the effective elastic modulus G'_{eff} is the function of deposition speed $\bar{V}$ and equilibrium elastic modulus G'_{eq}. Therefore, deposition speed $\bar{V}$ (2 ~ 10 mm/s), equilibrium shear modulus G'_{eq}

$(4 \times 10^4 \sim 1.5 \times 10^5$ Pa) span distance L $(0.5 \sim 2.5$ mm), filament diameter D $(\sim 0.2$ mm), and gel density ρ_{gel} $(\sim 4.1$ g/cm^3) become important parameters on δz_{mid}.

In the case of all these variables, dimensionless analysis will reduce the number of variables to a more tractable set of dimensionless groups to characterize interrelationships. The Buckingham Pi method [17] is a systematic procedure for finding characteristic dimensionless groups associated with particular problem and then discovering functional relationships between these numbers. Three elementary dimensions (i.e., mass, length, and time) are taken to implement Buckingham Pi theorem, and there should be 3 (= 6 − 3) independent dimensionless groups. Three dimensionless groups are then defined as follows:

$$\pi_1 = \frac{\delta z_{mid}}{D} = \frac{\text{Midpoint deflection}}{\text{Filament diameter}} \tag{3}$$

$$\pi_2 = \frac{L}{D} = \frac{\text{Filament length}}{\text{Filament diameter}} \tag{4}$$

$$\pi_3 = \frac{\rho_{gel} \bar{V}^2}{G'_{eq}} = \frac{\text{Motion energy}}{\text{Elastic potential energy}} \tag{5}$$

Although the dimensionless groups are not unique, most of them still have a specific and explicit physical meaning. Physically, the first and second group indicates the deflection, and length of spanning filaments with respect to the filament diameter. The third group can be regarded as the ratio of motion energy and elastic potential energy.

Relations of Dimensionless Groups

Based on the above dimensional argument, three relevant dimensionless groups can be written as:

$$\frac{\delta z_{mid}}{D} = f\left(\frac{L}{D}, \frac{\rho_{gel} \bar{V}^2}{G'_{eq}} \right) \tag{6}$$

where f is an as-yet undetermined correlating function, which is either monomial or non-monomial form [18]. There are various possible forms of three dimensionless groups' combination. Firstly, the monomial (i.e., power series) form of f function was tried as:

$$\frac{\delta z_{mid}}{D} = y_0 \left(\frac{L}{D} \right)^{y_1} \left(\frac{\rho_{gel} \bar{V}^2}{G'_{eq}} \right)^{y_2} \tag{7}$$

where y_0, y_1, and y_2 are constants to be determined.

Generally speaking, in order to seek the dependence of one dimensionless group upon the others, one must keep the remaining groups constant and plot the curve of one group versus the others. In order to investigate the effect of L/D and $\rho_{gel} \bar{V}^2/G'_{eq}$ on $\delta z_{mid}/D$, their values were plotted while keeping another group constant in Figure 2. In Figure 2 (a), the $\delta z_{mid}/D$ shows a monotone increasing power function trend with L/D at different constant $\rho_{gel} \bar{V}^2/G'_{eq}$. In Figure 2 (b), the same trend of $\delta z_{mid}/D$ as a function of $\rho_{gel} \bar{V}^2/G'_{eq}$ at varying constant L/D except for the existence of intercept when $\rho_{gel} \bar{V}^2/G'_{eq}$ approaches zero. Then, the monomial Eq. (7) was modified to non-monomial form by adding another undetermined constant y_3 as follows:

$$\frac{\delta z_{mid}}{D} = y_0 \left(\frac{L}{D} \right)^{y_1} \left(y_3 + \left(\frac{\rho_{gel} \bar{V}^2}{G'_{eq}} \right)^{y_2} \right) \tag{8}$$

The values of y_0, y_1, y_2, and y_3 were calculated from experimental data by using least squares regression, and listed in Table 1.

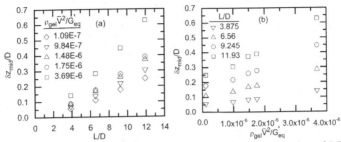

Figure 2 Relations between dimensionless groups (a) $\delta z_{mid}/D$ as a function of L/D by keeping $\rho_{gel}\bar{V}^2/G'_{eq}$ constant; (b) $\delta z_{mid}/D$ as a function of $\rho_{gel}\bar{V}^2/G'_{eq}$ by keeping L/D constant.

Table 1 Empirical constants obtained by least-square regression

y_0	y_1	y_2	y_3
1.24×10^5	1.33	1.28	7.19×10^{-8}

Validation and Prediction of Empirical Model

Figure 3 showed the comparison of empirical model predictions with previous experimental data. It is clear from the figure that the model is in good agreement with experimental data and the deviations between the empirical model predictions and experimental data mostly fall into ±10% error ranges. Figure 4 illustrated the predicted values of $\delta z_{mid}/D$ as a function of both L/D and $\rho_{gel}\bar{V}^2/G'_{eq}$.

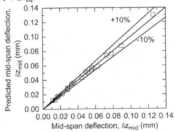

Figure 3 Comparison of model predictions with experimental data

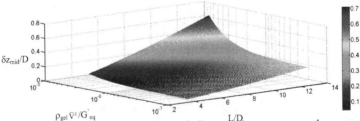

Figure 4 Model predictions as a function of dimensionless groups and compared with experimental data

MATHEMATICAL MODELING OF VISCOELASTIC CATENARY

Bending Moment of Viscoelastic Materials

The stress-strain behavior of a viscoelastic material under a constant tensile stress σ can be analogue to Kelvin-Voigt model with parallel combination of two ideal rheological elements, spring and dashpot, as illustrated in Figure 5. The spring represents the elastic behavior, and the dashpot represents the viscous behavior of the material. The traction coefficient of the dashpot λ is determined by the viscosity η with the relationship $\lambda = 3\eta$. The relationship between the tensile stress σ_e and the tensile strain ε_e in the spring element is given by Hooke's law as:

$$\sigma_e = E \cdot \varepsilon_e \tag{9}$$

The relationship between the traction force σ_v and the traction strain ε_v in the dashpot element can make an analogy to shear flow constitutive equation of Newtonian fluid as:

$$\sigma_v = \lambda \cdot (d\varepsilon_v / dt) \tag{10}$$

The total stress σ is the sum of elastic and viscous stress as:

$$\sigma = \sigma_e + \sigma_v = E \cdot \varepsilon_e + \lambda \cdot (d\varepsilon_v / dt) \tag{11}$$

Since the total strain ε is equal to the ε_e and ε_v as:

$$\varepsilon = \varepsilon_e = \varepsilon_v \tag{12}$$

Eq. (11) can be rewritten as:

$$\sigma = E \cdot \varepsilon + \lambda \cdot (d\varepsilon / dt) \tag{13}$$

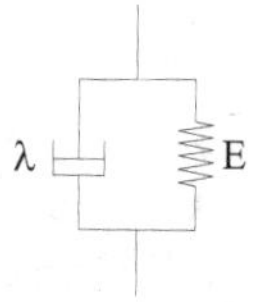

$$\lambda \; \boxed{} \quad \lessgtr E$$

Figure 5 Mechanical model analogs of Kelvin-Voigt viscoelastic materials

Assuming that a plane cross section normal to the axis of the filament remains plane after bending, it can be easily shown by simple bending theory that at any section of the filament, the longitudinal strain ε can be expressed as:

$$\varepsilon = Y / R \tag{14}$$

where Y is the distance of surface from neutral axis, and R is the radius of curvature of the filament. Here, we set θ (s, t) as the kinetic angle of the tangent to the centerline with the horizontal. By simply setting $\partial a/\partial b = a_b$, the curvature $d\theta/ds$ can be expressed as:

$$d\theta / ds = \theta_s = 1/R \tag{15}$$

Then, substituting Eq. (14) and (15) into Eq. (13), multiplying it by YdA, and integrating it over the section of the filament:

$$\int \sigma \cdot Y dA = E \cdot \theta_s \int Y^2 dA + \lambda \cdot \theta_{st} \int Y^2 dA \tag{16}$$

where $\int \sigma \cdot Y dA$ is the bending moment M and $\int Y^2 dA$ is the area moment of inertia I. Thus, the M of viscoelastic filament can be expressed as [19]:

$$M = EI \cdot \theta_s + \lambda I \cdot \theta_{st} \tag{17}$$

Viscoelastic Catenary Model Development

The planar forces and moments analysis of the viscoelastic filament element is shown in Figure 6. The filament is assumed to be under uniformly distributed load. The length of the element ds can be separated into horizontal component dx, and vertical component dy. The gravity force acting on the element can be expressed by the product of weight distribution w, and element length ds. At the lower end, there exist a tangential tension force T, a normal shear force N, and a counter clockwise direction bending moment M. At the upper end, there exist opposite direction tangential tension force T + dT, normal shear force N + dN, and bending moment M + dM, respectively. The bending slope at any position of the filament can be measured by slope angle θ. The derivation of viscoelastic catenary model follows the viscous catenary derived by Teichman et al. [20].

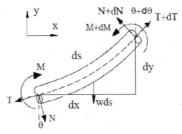

Figure 6 Forces and moments acting on an element of the filament

Horizontal forces balance: $\sum F_H = 0$

$$\frac{d}{ds}(T \cdot \cos\theta - N \cdot \sin\theta) = 0 \tag{18}$$

$$\Rightarrow T_s \cdot \cos\theta - T \cdot \sin\theta \cdot \theta_s - N_s \cdot \sin\theta - N \cdot \cos\theta \cdot \theta_s = 0$$

Vertical forces balance: $\sum F_V = 0$

$$\frac{d}{ds}(T \cdot \sin\theta - N \cdot \cos\theta) = 0 \tag{19}$$

$$\Rightarrow T_s \cdot \sin\theta + T \cdot \cos\theta \cdot \theta_s + N_s \cdot \cos\theta - N \cdot \sin\theta \cdot \theta_s = w$$

The tension force T, and shear force N in Eq. (18) and (19) can be expressed respectively as [21]:

$$T = \frac{w \cdot \cos\theta - N_s}{\theta_s} \tag{20}$$

$$N = \frac{T_s - w \cdot \sin\theta}{\theta_s} \tag{21}$$

By eliminating the tension force T in Eq. (20) and (21), we can get

$$2w \cdot \sin\theta + N \cdot \theta_s + \frac{N_{ss}}{\theta_s} + \frac{w \cdot \cos\theta}{\theta_s^2} - \frac{N_s \cdot \theta_{ss}}{\theta_s^2} = 0 \tag{22}$$

The universal relationship between shear force N and bending moment M has the form:

$$M_s + N = 0 \tag{23}$$

And previous derivation has showed the bending moment of viscoelastic filament in Eq. (17), so the Eq. (23) can be rewritten as:

$$N = -M_s = -(EI \cdot \theta_{ss} + \lambda I \cdot \theta_{sst}) \tag{24}$$

Substituting Eq. (24) into Eq. (22), multiplying by $\cos\theta$, and integrating it, we can get

$$EI\left(\frac{\theta_{sss}}{\theta_s}\cdot\cos\theta+\theta_{ss}\cdot\sin\theta\right)+\lambda I\left(\frac{\theta_{ssst}}{\theta_s}\cdot\cos\theta+\theta_{sst}\cdot\sin\theta\right)+w\cdot\frac{\cos^2\theta}{\theta_s}=f_1(t) \tag{25}$$

where $f_1(t)$ is a constant of integration. By setting $\theta=0$, Eq. (25) can be turned into

$$EI\cdot\frac{\theta_{sss}}{\theta_s}+\lambda I\cdot\frac{\theta_{ssst}}{\theta_s}+\frac{w}{\theta_s}=f_1(t) \tag{26}$$

Substituting Eq. (24) into Eq. (20), the result can be written as:

$$EI\cdot\frac{\theta_{sss}}{\theta_s}+\lambda I\cdot\frac{\theta_{ssst}}{\theta_s}+\frac{w}{\theta_s}=T(x,t) \tag{27}$$

Comparing Eq. (26) with (27), using the setting condition $\theta=0$, we can get

$$f_1(t)=T(x,t)=T(0,t) \tag{28}$$

That means the tension force $T(x, t)$ is only a time dependent variable as:

$$T_x=0;\ T(x,t)=T(t) \tag{29}$$

Substituting Eq. (28), and (29) into Eq. (25), it turned into

$$EI\left(\frac{\theta_{sss}}{\theta_s}\cdot\cos\theta+\theta_{ss}\cdot\sin\theta\right)+\lambda I\left(\frac{\theta_{ssst}}{\theta_s}\cdot\cos\theta+\theta_{sst}\cdot\sin\theta\right)+w\cdot\frac{\cos^2\theta}{\theta_s}=T(t) \tag{30}$$

Multiplying Eq. (30) by $\theta_s/\cos^2\theta$ and integrating it, we can get

$$(EI\cdot\theta_{ss}+\lambda I\cdot\theta_{sst})\cdot\sec\theta+ws=T(t)\cdot\tan\theta+f_2(t) \tag{31}$$

where $f_2(t)$ is a constant of integration.

The symmetry determines that $\theta(0,t)=\theta_{ss}(0,t)=\theta_{sst}(0,t)=0$, that means $f_2(t)=0$. Besides, the boundary conditions at the supported ends are $\theta(\pm L/2,t)=0$. Thus, Eq. (31) can be formulated as:

$$(EI\cdot\theta_{ss}+\lambda I\cdot\theta_{sst})\cdot\sec\theta+ws=T(t)\cdot\tan\theta \tag{32}$$

In order to figure out the time-dependent expression for tension force $T(t)$, we need to analyze the filament movement dynamics during shape evolution. Figure 7 shows the element movement during the filament sagging process. Setting $u(x, t)$ is the horizontal displacement, and $v(x, t)$ is the vertical displacement of a cross-section element at location x.

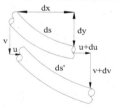

Figure 7 Time dependent displacement of an element of the filament [22].

If ds is the original length of the element, and ds' is its new length, they can be expressed respectively as:

$$ds^2=dx^2+dy^2 \tag{33}$$

$$ds'^2=(dx+du)^2+(dy+dv)^2 \tag{34}$$

For small to moderate deflection, $\theta<1$, $s\approx x$, so that $v_x\approx\theta$. Corrected to the second order of small quantities, the filaments fractional change in length can be expressed as:

$$\left(\frac{ds'-ds}{ds}\right) = u_x + \frac{v_x^2}{2} \tag{35}$$

Then, the tension $T(t)$ in the viscoelastic filament can be expressed as:

$$T(t) = EA\left(\frac{ds'-ds}{ds}\right) + \lambda A\left(\frac{ds'-ds}{ds}\right)_t = EA\left(u_x + \frac{v_x^2}{2}\right) + \lambda A\left(u_{xt} + v_x v_{xt}\right) \tag{36}$$

Integrating Eq. (36) by using the boundary condition $u_t(\pm L/2, t) = 0$, and the symmetry of the problem which leads to $u_t(0, t) = 0$, we can get

$$\int_0^{L/2} T(t) dx = (L/2)T(t) = EA\int_0^{L/2}\frac{v_x^2}{2}dx + \lambda A\int_0^{L/2}\left(\frac{v_x^2}{2}\right)_t dx \Rightarrow \tag{37}$$

$$T(t) = \frac{EA}{L}\int_0^{L/2}\theta^2 dx + \frac{\lambda A}{L}\int_0^{L/2}\left(\theta^2\right)_t dx$$

By substituting Eq. (37) into Eq. (32), the final equation of filament dynamic catenary with boundary condition can be written as:

$$\begin{cases}\left(EI\cdot\theta_{xx} + \lambda I\cdot\theta_{xxt}\right) - \theta\cdot\left(\frac{EA}{L}\int_0^{L/2}\theta^2 dx + \frac{\lambda A}{L}\int_0^{L/2}\left(\theta^2\right)_t dx\right) = -wx \\ \theta(0, t) = \theta(\pm L/2, t) = 0; \theta(x, 0) = 0\end{cases} \tag{38}$$

The first term on the left hand of Eq. (38) represents the resistance to viscoelastic bending effect; and the second term represents the resistance to viscoelastic stretching effect; the inhomogeneous force term on the right hand arises from the weight of the filament.

Analytical Method for Model Solution

During the early stage of deflection, $\theta \ll 1$, the second term on the left hand of Eq. (38) can be neglected, and it can be simplified as:

$$EI\cdot\theta_{xx} + \lambda I\cdot\theta_{xxt} = -wx \tag{39}$$

Integrating Eq. (39) with the initial condition $\theta(0, t) = 0$, $\theta(\pm L/2, t) = 0$, and $\theta(x, 0) = 0$ yields

$$\theta(x, t) = \left(-\frac{wx^3}{6EI} + \frac{wL^2x}{24EI}\right)\left[1 - \exp\left(-\int_0^t\frac{E}{\lambda}dt\right)\right] \tag{40}$$

The vertical displacement $v(x, t)$ is

$$v(x, t) = \int_{-L/2}^{} \theta dx = -\frac{w}{24EI}\left[\left(\frac{L}{2}\right)^2 - x^2\right]^2\left[1 - \exp\left(-\int_0^t\frac{E}{\lambda}dt\right)\right] \tag{41}$$

This solution is resemble to the elastic beam deflection expression, and describes the initial bending deflection of whole filament in very short time. As time progresses, the stretching effect starts to dominate the deflection process from the mid-span point to lateral supported ends. And the bending effect will be localized to a $\delta(t)$ neighborhood of the lateral attachment boundaries, while elsewhere the filament is subject primarily to stretching strain without cross-section shrinkage. Thus, over most of the filament we have a balance between stretching and gravity for the stretching solution, which satisfies:

$$\begin{cases}\theta\cdot\left[\frac{EA}{L}\int_0^{L/2}\theta^2 dx + \frac{\lambda A}{L}\int_0^{L/2}\left(\theta^2\right)_t dx\right] = wx \\ x \in\left[-L/2 + \delta, L/2 - \delta\right]\end{cases} \tag{42}$$

Squaring Eq. (42), and integrating it, we can get

$$\int_0^{L/2} \theta^2 dx = \frac{w^2 L^5}{24A^2} \left[E \int_0^{L/2} \theta^2 dx + \lambda \int_0^{L/2} \left(\theta^2 \right)_t dx \right]^{-2} = \frac{L^3 \theta^2}{24x^2} \tag{43}$$

Similarly, we can also integrate $\left(\theta^2 \right)_t$ to get

$$\int_0^{L/2} \left(\theta^2 \right)_t dx = \frac{L^3 \left(\theta^2 \right)_t}{24x^2} \tag{44}$$

Substituting Eq. (43) and (44) into Eq. (42), we can get

$$E\theta^3 + \frac{2}{3} \lambda \left(\theta^3 \right)_t = \frac{24wx^3}{AL^2} \tag{45}$$

Integrating Eq. (45), it yields

$$\theta(x,t) = x \left[\frac{24w}{EAL^2} \left(1 - \exp\left(-\int_0^t \frac{3E}{2\lambda} dt \right) \right) \right]^{1/3} \tag{46}$$

The vertical displacement $v(x, t)$ is

$$v(x,t) = \int_{-L/2}^x \theta dx = \left(\frac{x^2}{2} - \frac{L^2}{8} \right) \left[\frac{24w}{EAL^2} \left(1 - \exp\left(-\int_0^t \frac{3E}{2\lambda} dt \right) \right) \right]^{1/3} \tag{47}$$

In lateral boundaries, the filament must be highly curved, and the stretching solution need to change rapidly to match the supported end condition, $\theta(\pm L/2, t) = 0$. By balancing the bending term in Eq. (38), $EI \cdot \theta_{xx} + \lambda I \cdot \theta_{xxt} \sim ED^4\theta/\delta^2 + \lambda D^4\theta/\left(\delta^2 t \right)$ with the weight wx, and substituting stretching solution Eq. (32), yields the scaling width $\delta(t)$ of the bending boundary layer,

$$\delta \sim \left(E^{2/3} + \lambda E^{-1/3} / t \right)^{1/2} \left(D^3 / wL \right)^{1/3} \left[1 - \exp\left(-\int_0^t (3E/2\lambda) dt \right) \right]^{1/6} .$$

Equilibrium Shape Profiles of Viscoelastic Spanning Elements

As the time $t \to \infty$, the spanning filament reached its equilibrium shape, and the equilibrium width $\delta(t \to \infty)$ of the bending boundary layer can be scaled as $\delta \sim \left(ED^5 / wL \right)^{1/3}$. In the lateral bending boundary layer, the equilibrium shape was calculated through the bending solution Eq. (41); while in the middle stretching layer, the equilibrium shape was calculated through the stretching solution Eq. (47).

$$\begin{cases} v(x, t \to \infty) = -\frac{w}{24EI} \left[\left(\frac{L}{2} \right)^2 - x^2 \right]; x \in [-L/2, -L/2+\delta] \cap [L/2-\delta, L/2] \\ v(x, t \to \infty) = \left(\frac{x^2}{2} - \frac{L^2}{8} \right) \left(\frac{24w}{EAL^2} \right)^{1/3}; x \in [-L/2+\delta, L/2-\delta] \end{cases} \tag{48}$$

From Eq. (48), we can see that the final shape of the spanning filament is determined by its weight distribution w, diameter D, span distance L, and elastic modulus G'. For a specific colloidal ink deposited by the same tip, only span distance L, and elastic modulus G' are variables for this model equation. Previous research [23] has correlated the G' to the shear history. Here, we assume that the proper effective elastic modulus G'eff needs to be selected and predetermined to fit the data.

Figure 8 shows the equilibrium deflection profile of selected filaments at different pH values. We can see that as the pH value decreases, the corresponding deflection decreases remarkably. This phenomenon is the result of the gel strength reduction with the pH decrease.

Besides, the span distance augment leads to the increase of filaments deflection profile at all pH values. Here, we use a piecewise function of Eq. (48) to describe the deflection profile in the whole intervals of span distance. Although this method will lead to the unsmooth of the fitting curve and inaccurate in the lateral range, the fitting curve still can provide precise prediction in the middle range and boundary layer near two support ends.

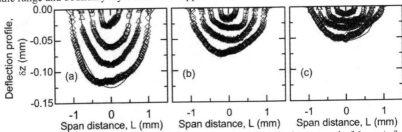

Figure 8 Equilibrium deflection profiles of spanning filaments deposited at a speed of 6 mm/s for span distances L = 0.775 ($\square$), 1.312 ($\circ$), 1.849 ($\triangle$), and 2.386 ($\diamond$) mm from PZT colloidal inks at varying pH = (a) 7.60, (b) 6.85, and (c) 6.15.

Figure 9 shows the varying trend of the G'_{eff} with the pH value of PZT colloidal inks at the same deposition speed of 6 mm/s. As the pH value increases, the gel strength is attenuated, so that the corresponding G'_{eq} decreases simultaneously. Due to the shear flow during extrusion process, the G'_{eff} is always below the G'_{eq} for PZT colloidal inks with different gel strength. And the G'_{eff} also decreases with the descending of the gel strength caused by the pH reduction.

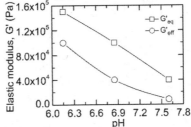

Figure 9 Comparison of effective elastic modulus G'_{eff} and equilibrium elastic modulus G'_{eq} of PZT colloidal inks as a function of pH at a deposition speed of 6 mm/s.

Figure 10 shows the deposition speed's influence on the equilibrium deflection profiles for PZT colloidal inks at pH = 6.15. We can find that the higher deposition can generate larger deflection profile for various span distances. This can be explained by the relationship between the deposition speed and G'_{eff}. Here the pH is a constant of 6.15, the only factor can alter the G'_{eff} is the deposition.

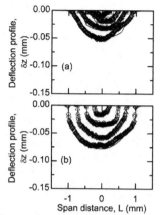

Figure 10 Equilibrium deflection profiles of spanning filaments deposited at speeds of (a) 2mm/s, (b) 8mm/s, for span distances L = 0.775 (□), 1.312 (○), 1.849 (Δ), and 2.386 (◇) mm from PZT colloidal inks at pH = 6.15.

According to the selected values of G'_{eff}, we can know the G'_{eff} decreases as the deposition speed increases, as illustrated in Figure 11. In Figure 11, the G'_{eff} varying trend of PZT colloidal inks at pH = 6.15 was illustrated as a function of deposition speeds. Comparing to the G'_{eq} (dash-dot line), the G'_{eff} get close to the G'_{eq} at lower deposition speed, and as the deposition speed increases, the G'_{eff} decreases from slow to dramatically. This trend indicates that the G'_{eff} is sensitive to the deposition speed, whose increase can greatly attenuate the G'_{eff}.

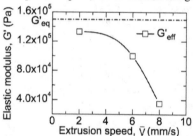

Figure 11 The effective elastic modulus G'_{eff} of PZT colloidal inks at pH = 6.15 vary with deposition speeds.

Time Dependent Shape Evolution of Viscoelastic Spanning Filaments

Since the initial bending is in very short time, we just use the stretching function of Eq. (47) to track the mid-span dynamic deflection by setting x = 0 as:

$$v(0, t) = -\frac{1}{4}\left[\frac{3wL^4}{EA}\left(1 - \exp\left(-\int_0^t \frac{3E}{2\lambda}dt\right)\right)\right]^{1/3} \tag{49}$$

Although Eq. (49) is not accurate enough to describe the initial deflection, it still shows the whole deflection dynamics of mid-span point. From Eq. (49) we can see that except for the G'_{eff}, the ratio of Young's modulus and traction coefficient E/λ is another important factor to control the deflection velocity.

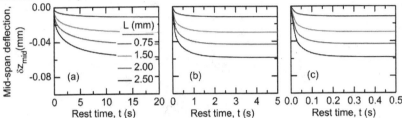

Figure 12 Mid-span point dynamic deflection of spanning filaments from PZT colloidal inks at pH = 6.15 by assuming $G'_{eff} = 10^5 Pa$; and $E/\lambda = $ (a) 0.1, (b) 1.0, and (c) 10.

As we can see from Figure 12, the increase of span distance results in the increase of deflection at any instant. The deflection velocity exponentially decreases with the rest time. As the E/λ increases, the time for mid-span to reach its equilibrium state decreases dramatically. It can be concluded that the viscosity of the ink after deposition can significantly change the deflection velocity.

CONCLUSIONS

In this work, an empirical model was used to characterize the mid-span deflection of spanning filament by using dimensional analysis method. In the range of experimental data, this model demonstrated excellent predictions. Except for this blackbox model, a viscoelastic catenary model has been developed to describe the time dependent deflection of spanning filaments. The simulation results indicated that this model provides good predictive capability for mesoscale spanning filaments. Although the current work is limited to colloidal gels, the knowledge gained here may be easily extended to other complex ink systems such as partially melted thermoplastic polymers and metals. The contributions made by this research will open new pathways to serve as guidelines for new inks designs and 3D shape evolution control.

Further refinements in process modeling are needed to characterize the ink flow inside the deposition nozzle and the structural recovery immediately after deposition. Dynamic measurements of spanning filaments deflection are also needed to be carried out to validate the time dependent behavior of this model. These analyses will provide important information for the design of future inks and help define the processing parameters. High-performance inks combined with accurate process modeling of the direct-write assembly technique will enable the creation of microvascular networks with unparalleled complexity and commercialization of technological applications in biomedical, advanced materials and micro-fluidics.

REFERENCES
[1]Chrisey, D.B., *The Power of Direct Writing*. Science, 2000. **289**: p. 879-81.
[2]Pigue, A. and D.B. Chrisey, *Direct Write Technologies for Rapid Prototyping Applications: In Sensors, Electronics and Integrated Power Sources* 2002, San Diego, CA: Academic Press.
[3]Lewis, J.A., *Direct ink writing of 3D functional materials*. Advanced Functional Materials, 2006. **16**: p. 2193-04.
[4]Lewis, J.A. and G.M. Gratson, *Direct writing in three dimensions*. Materials Today, 2004. **7**: p. 32-39.

[5]Smay, J.E., J.C. III, and J.A. Lewis, *Colloidal Inks for Directed Assembly of 3-D Periodic Structures.* Langmuir, 2002. **18**: p. 5249-37.

[6]Joannopoulos, J.D., P.R. Villeneuve, and S.H. Fan, *Photonic crystals: putting a new twist on light.* Nature, 1997. **386**: p. 143-49.

[7]Blaaderen, A.v., R. Ruel, and P. Wiltzius, *Template-Directed Colloidal Crystallization.* Nature, 1997. **385**: p. 321-324.

[8]Therriault, D., S.R. White, and J.A. Lewis, *Chaotic mixing in three-dimensional microvascular networks fabricated by direct-write assembly.* Nature Materials, 2003. **2**(4): p. 265-71.

[9]Bruneaux, J., D. Therriault, and M.-C. Heuzey, *Micro-Extrusion of Organic Inks for Direct-Write Assembly.* Journal of Micromechanics and Microengineering, 2008. **18**: p. 115020 (11pp).

[10]Michna, S., W. Wu, and J.A. Lewis, *Concentrated Hydroxyappatite Inks for Direct-Write Assembly of 3-D Periodic Scaffolds.* Biomaterials, 2005. **26**: p. 5632-9.

[11]Smay, J.E., et al., *Directed Colloidal Assembly of 3D Periodic Structures.* Advanced Materials, 2002. **14**: p. 1279-83.

[12]Lewis, J.A., *Direct Ink Writing of Three-Dimensional Ceramic Structures.* Journal of the American Ceramic Society, 2006. **89**(12): p. 3599-609.

[13]Therriault, D., et al., *Fugitive Inks for Direct-Write Assembly of Three-Dimensional Microvascular Network.* Advanced Materials, 2005. **17**(4): p. 395-9.

[14]Therriault, D., S.R. White, and J.A. Lewis, *Rheological Behavior of Fugitive Organic Inks for Direct-Write Assembly.* Applied Rheology, 2007. **17**(1): p. 10112-1-8.

[15]Shigley, J.E. and C.R. Misckhe, *Mechanical Engineering Design.* 5th ed. 1989.

[16]Channell, G. and C. Zukoski, *Shear and Compressive Rheology of Aggregated Alumina Suspensions.* AICHE Journal, 1997. **43**(7): p. 1700-8.

[17]Munson, B.R., D.F. Young, and T.H. Okiishi, *Fundamentals of Fluid Mechanics.* 2002, New York: John Wiley & Sons, Inc.

[18]Szirtes, T., *Applied Dimensional Analysis and Modeling.* 1998, New York: McGraw-Hill.

[19]Lenczner, D., *Deflection of Beams and Composite Walls Subject to Creep* Building Science, 1971. **6**: p. 45-51.

[20]Teichman, J. and L. Mahadevan, *The Viscous Catenary.* Journal of Fluid Mechanics, 2003. **478**: p. 71-80.

[21]Love, A.E.H., *A Treatise on the Mathematical Theory of Elasticity.* 1944: Dover.

[22]Irvine, H.M., *Cable Structures.* 1981, New York: Dover Publication, Inc.

[23]Nadkarni, S.S. and J.E. Smay, *Concentrated Barium Titanate Colloidal Prepared by Bridging Flocculation for Use in Solid Freeform Fabrication.* Journal of the American Ceramic Society, 2006. **89**(1): p. 96-103.

Composites Processing

COLLOIDAL PROCESSING OF CERAMIC-CERAMIC AND CERAMIC-METAL COMPOSITES

Rodrigo Moreno
Instituto de Cerámica y Vidrio, CSIC
Kelsen 5, 28049 Madrid, Spain.

ABSTRACT

Colloidal processing has demonstrated its suitability to produce complex-shaped ceramic parts with tailored microstructure. Less effort has been devoted toward processing of metal powders and ceramic-metal composites because of the high density and complex surface behaviour of metal powders. By combining different shaping methods it is possible to produce complex 3-D bodies as well as single or multilayer coatings, self-sustaining films and laminates In this work, the colloidal processing of Ni powders and Ni-Al$_2$O$_3$ and Ni-ZrO$_2$ composites is studied considering the rheological behavior of the suspensions and specific processing parameters that allow manufacturing complex 2-D and 3-D structures with layers having controlled thickness. Innovative processing strategies resulting from the combination of conventional consolidation mechanisms and shaping routes are explored, such as the production of bulk bodies and layered materials by slip casting of metal and ceramic mixtures, the fabrication of substrates by tape casting, and the application of coatings by dipping.

INTRODUCTION

The continuous progress of emerging technologies has promoted the development of new engineered materials with enhanced properties capable to meet their increasing demands. A first issue to be considered before selecting a specific material is the profile of properties it must satisfy. A second aspect is the capability to produce a component with the desired geometry, size and microstructure for its use in the corresponding device or component. This implies that the selection of materials and processing routes are intimately related. However, the shaping itself can change or even determine the properties of the final material.

It is customary to classify the different engineering materials into six broad families: metals, polymers, elastomers, ceramics, glasses and hybrids—composite materials made by combining two or more of the others.[1] Other authors reduce the number of families to four: metals, polymers (including elastomers), ceramics (including glasses), and composites.[2]

Metals are usually stiff and tough. In general, metals are soft and easily deformed, but they can be made strong by alloying and by mechanical and heat treatment.[3] The ductility of metals allows them to be shaped by deformation processes. The main limitations of metals are their reactivity and the low resistance to corrosion. Ceramics are non-metallic, inorganic solids, that harden after a thermal treatment that provides high resistance to corrosion and chemicals, high refractoriness, hardness, etc. However, a major limitation of ceramic materials is their inherent brittleness due to the high directionality of the covalent bond. This gives low tolerance to stress concentrations thus difficulting processing over that of metals. Since ceramics are usually formed from powders some typical shaping processes of metals like deformation methods cannot be used for ceramics. An important drawback of powder processing is the strong tendency of powders to agglomeration, which is one of the most important parameters to be controlled in processing science and technology.

POWDER PROCESSING

In contrast with polymers, a common issue of ceramics and metals is that both types of materials can be processed using powder processing techniques, consisting of compaction of powders and further heat treatment.[4,5] Dry processing is mostly used among pulvimetallurgy techniques due to the surface dissolution and the high density of metal powders, which difficult wet processing.

However, ceramics are often processed in the wet state by using well-dispersed suspensions. If sedimentation is retarded and dissolution and surface reactions are controlled, metals can be processed in water as ceramic powders. This is possible because controlled oxidation at the particles surface allows the metal particle to have a thin oxidized layer that behaves as a ceramic oxide. If the thickness of this oxidized layer is sufficiently small it can preserve the particle to further oxidize and can be reduced to the metal state by heating in a controlled atmosphere.

The major weakness of ceramics continues to be brittleness, since the partial covalency of the bonds makes them very rigid and does not allow plasticity.[6] The mechanical strength of a material is directly dependen on its critical stress intensity factor or toughness and is inversely proportional to the square root of the critical flaw size. This constitutes the basis for the establishment of the proper relationships between processing and properties. Better strength can be obtained in two ways: on one hand, by increasing the toughness of the material, which is possible by selecting better new materials or by enhancement of the properties through a suitable microstructural design; on the other hand, the strength increases when the critical flaw size decreases, that is, by controlling the different steps involved in the processing cycle.

The reinforcement of ceramic-based composites can be achieved by different mechanisms, the most frequently employed being the addition of secondary phases that disturb the crack path during its propagation, or the design of tailored microstructures with well defined interfaces that makes the crack to deviate, arrest or bifurcate.[6] To take maximum profit of the second phase as a reinforcing mechanism, processing is of critical importance in assuring the optimal dispersion inside the matrix. Similarly, processing must allow the control of the layers composition, thickness and shape, in the case of laminates, where the existance of sharp, well-defined interfaces is a critical issue. The presence of uncontrolled defects, such as the bubbles appearing in wet forming processes, can lead to big defects that destroy the material structure and properties. The defect minimization has to be so effective that defects above a certain size never occur. In order to understand how this control of microstructure is possible it is necessary to know the fundamentals of shape forming of pieces from powders.

A composite can be defined as a multiphase material that maintains, at least partially, the properties of the different constituents such that a better combination of properties is obtained.[7,8] For engineering applications composites are specifically tailored and made, but there are many composites naturally occurring in nature, such as moluscs shells, wood, and bones, among others.

In the simplest case composites consist on two phases; a major, continuos one that is called the *matrix*, and a minor phase, often called the *dispersed phase*. Materials having a ceramic matrix are known as ceramic matrix composites (CMC). Composites can be divided into three main groups:[7] particle-reinforced, fiber-reinforced, and structural composites. The flow diagram of figure 1 summarizes the classification of composites including some special cases such as platelets, or coatings and functionally graded materials as particular examples of structural composites. In the first case, particles are equiaxed (e.g. spherical) or have a low shape factor when particles are non-spherical.

Ceramics are mainly produced by powder processing techniques, according to the following processing steps: 1) powder synthesis and/or preparation for further consolidation, 2) consolidation of powders into a self-supported shaped body, the so-called *green body*, 3) drying and burn out of organics, 4) sintering at high temperature to reach the final microstructure and properties, and 5) final machining and shaping, which is the most expensive step due to the hardness of ceramics. Maximum control at any stage is necessary since defects introduced in one step are very difficult to remove and will persist in the next steps. If the starting powders have defects they coould not be removed during consolidation, so that the purity and control of the starting powders is a first key parameter to obtain defect-free materials.[9,10] Figure 2 summarizes the main steps of a ceramic processing cycle.

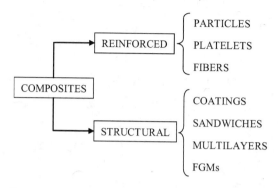

Figure 1. Schematic classification of ceramic-based composites.

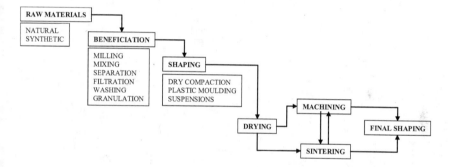

Figure 2. Flow-chart of the different steps of a ceramic processing cycle.

The production of a ceramic material starts with the selection of suitable raw materials, which can be natural or synthetic. Anyway, the powders need to be conditioned for further processing operations, through the so-called beneficiation processes.[5,11,12] These include any transformation of the physicochemical properties of the starting powder such as milling, mixing, separation, washing, granulation, etc. The powder is then consolidated into a part with defined size and shape by any of the shape forming methods available in ceramic technology. Forming methods can be classified according to three categories, depending on the relative content of liquid characteristic of the process: 1) dry pressing methods, 2) plastic forming, and 3) colloidal shaping methods that make use of suspensions.

The transformation steps occurring during shaping are commonly performed with the help of processing aids that allow the control of the characteristics of the powders at the different stages. These processing aids can be classified as organic and inorganic additives.[13-16] The literature of processing science and technology is somewhat confusing as many types of additives have been identified for specific functions. In the simplest approach, there would be two wide types of additives according to the general function they have: dispersing additives and binding additives. The first group includes liquid media, deflocculants (with the different types and structures, polyelectrolytes, surfactants and wetting agents, salts, acids and bases, etc) and the second includes binders, plasticizers, thickeners, gel formers, coagulants, etc). A summary of the general types of additives used in ceramic processing is presented in Table 1.

Table 1. Types of additives used in ceramic processing.

Type of additve	Inorganic	Organic
Deflocculants	• Potential determining ions (pH) • Electrolytes • Salts (NaCO3,,Na2SiO4,, Tripolyphosphates, Hexametaphosphates, sulfonates	• Surfactants (non-ionic, anionic, cationic, zwitterionic) • Steric stabilizers (polymers) • Electrosteric stabilizers (polyelectrolytes; cationic, PEI; anionic, PAA)
Binders	• Colloidal silica • Clays • Silicates • Cements	• Cellulose derivatives (MC, CMC, HEC) • Poly(vinyl alcohol) • Poly(vinyl butyral) • Alginates, Starches, Dextrines, Gums
Plasticizers	Clays, bentonites	Glycols, Phtalates
Lubricants	Colloidal talc, Colloidal graphite	Fatty acids, Oils, Stearates
Coagulants	$CaCl_2$, $AlCl_3$, $CaCO_3$	Alginates, Urea
Gel formers		Alginates, Methylcellulose, Starches Agar, Agarose, Carrageenan, Gelatin

COLLOIDAL PROCESSING

If the moisture content increases over 50 vol% then suspensions are formed in which particles are dispersed in a liquid (normally water). The control of the interparticle forces allow one to produce stable suspensions in which particles are repelled each other and this repulsion maintains even during consolidation. This means that very homogeneous and dense green bodies can be obtained at very low cost and practically no investment. The reference technique for suspensions forming is slip casting, which is the source of a number of forming methods, like those based in filtration like pressure casting or centrifugal casting, and other methods based in consolidation mechanisms different to filtration, such as evaporation, flocculation, polymerization and gelation, among others.

A colloidal dispersion is defined as a multi-phase system in which one phase (or more) is dispersed in a continuous one or medium. At least one dimension lies within the nanometre (10^{-9} m) to micrometre (10^{-6} m) range, so that colloidal dispersions are mainly systems containing large molecules and/or small particles. The main factor determining the properties of a colloidal system are the particle size and shape, the surface properties, the interparticle interactions and the interactions between particles and dispersing medium. The interface between the dispersed phase and the dispersing medium plays an essential role in the surface properties, including adsorption, surface charge, electrical double layer, etc. There are many reference textbooks on colloid and surface science where the reader can find complementary information.[17-20]

In order to avoid agglomeration, the most important feature is to provide a good stabilization to the suspension. Deflocculants are used to provide repulsion among particles. This can be achieved by three general mechanisms: 1) electrostatic charging in the presence of electrolytes that promotes electrostatic repulsion, 2) adsorption of polymers that provide steric hindrance and impedes particles to touch each other, 3) adsorption of charged polymers (polyelectrolytes) where a steric hindrance avoids the contact at near-to-contact distance, whereas the charges promote an electrostatic repulsion at larger separation distances. These three types of stabilizing mechanisms are illustrated in figure 3.

Electrostatic

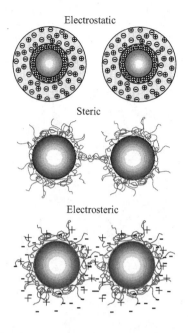

Steric

Electrosteric

Figure 3. Schematic representation of the electrostatic, steric and electrosteric mechanisms for the stabilization of particles in suspension.

When particles are immersed in a polar liquid, an electric double layer forms around the particles as a consequence of the acid-base reactions that take place in the medium. The double layer consists on a rigid layer of ions with opposite charge (counterions) to that of the particle surface strongly adsorbed to it and a diffuse layer in which the concentrations of counterions decreases as the separation distance increases. Acidic species like MOH_2^+ are formed at acidic conditions, whereas species MO^- are formed at basic conditions. There is a pH value at which the activities of the positive and the negative species at the particle surface are equal and the net charge density is zero. This pH defines the point of zero charge (PZC). The parameter used to measure the stability of a suspension is the zeta potential, which is the potential existing at the shear plane between the rigid and the diffuse layers. In order to assure the stability of a suspension it is necessary to have high values of zeta potential, so that working pH will be far away from the ZPC.

For particles of the same nature stable suspensions are reached when zeta potentials are typically above 25-30 mV, regardless the surface sign. However, when different types of particles are present, each one can have a different ZPC. There is a pH range where one type of particles is positively charged and the other one is negatively charged, so that electrostatic attraction occur leading to the so-called heterocoagulation phenomenom. This may cause agglomeration for particles with similar sizes, but can be a suitable route to produce composites by a core-shell route if one type of particles is at least one order of magnitude lower in size than the other type. A typical example for this is the manufacture of alumina-mullite composites by reaction sintering of alumina and silica, as illustrated in figure 4 for a mixture of submicronic alumina and colloidal silica. The finer silica particles surround the bigger alumina ones and reaction occurs on sintering.[21]

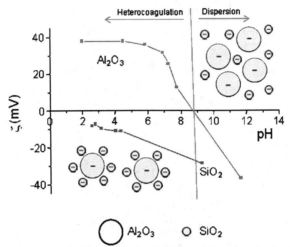

Figure 4. Zeta potentials of submicronsized alumina (large spheres) and nanosized silica (small spheres) suspensions with different size and interaction by dispersion or heterocoagulation depending on pH conditions.[21]

SHAPING OF CERAMICS AND COMPOSITES

Shaping refers to the consolidation of powders to obtain a green compact, after liquid removal. Then, it is possible to classify the general types of suspension forming techniques as a function of the consolidation mechanisms, as summarized in Table 2.[22]

Table 2. Suspension forming techniques as a function of consolidation mechanisms.

Filtration	Floculation-Coagulation	Gelation	Deposition-Evaporation
Slip casting	Short range forces	Gelcasting	Screen printing
- pressure	Temperature induced	Injection molding	Tape casting
- vaccuum	Coagulation casting	Thermogelation	Electrophoresis
- centrifugal	Direct solidification	Protein casting	CVD, PVD...
- microwaves		Starch consolidation	Dipping
Filter-pressing		Freeze casting	Spin coating
			Spraying

From this variety of shaping techniques and consolidation mechanisms, the most usual is the slip casting process, which is based on a filtration mechanism, that is, a separation of powders from dispersing liquid through a permeable mould that allows liquid filtration retaining the particles that adhere to the mould walls. The filtration process can be accelerated by external forces, like pressure, vacuum, etc. This process has the advantage that filtration is slow and gives time to the particles to rearrange into dense packing green structures. Then, consolidation is simultaneous to shaping and liquid loss. This is a fundamental difference with most of the other processes, in which the suspension is suddenly consolidated into a body by fast gelation, polymerization or freezing, but all the liquid is retained in the shaped part and must be removed later in a second stage. This leads to a worse dimensional control as a drying shrinkage occurs after consolidation and can even produce cracks or other defects.

The slip casting process consists on the preparation of a well-dispersed suspension with moderate to high solids loadings (typically between 30-40 vol% solids) to maintain viscosity low enough to be easily handled. The fluid suspension is then poured into the cavity of a permeable mould that leaves the liquid to pass through and impedes the filtration of the particles, which attach to the mould walls forming a wet cake that partially shrinks on drying. This is sufficient to allow the cast body to be removed from the mould.

The wall thickness (L) formation follows a parabolic growth with casting time (t), so that it is possible to control the casting time necessary to produce the desired thickness according to the simple relation $L^2=kt$, k being a constant. According to this, the thickness growth is limited and it is typically of a few milimeters, this meaning that slip casting is not suitable for thick pieces. The possibility of producing pieces with controlled thickness allows one to design laminates, coatings and functionally graded materials. The design of multilayers is done by casting specimens at different times and measuring the wall thickness of the greens; after checking and adjusting the thickness with observations at the scanning electron microscopy, the casting time needed to produce a certain thickness can be predicted with an error lower than 10%. Figure 5 shows the shape of a wall thickness formation curve. In order to obtain the same thickness of layers the casting time must increase progresively.

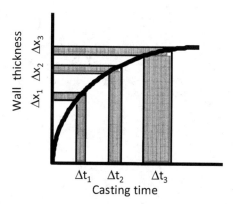

Figure 5. Design of multilayers and FGMs by sequential slip casting from the casting kinetics.

Simple control of the casting kinetics allows one to produce tailored microarchitectures with the desired layers distribution and thickness.[23,24] As an example, figure 6 shows the SEM microstructure of a multilayer of Al$_2$O$_3$/Al$_2$O$_3$-YTZP material (left) and a functionally graded material formed by discrete layers with changing compositon from YTZP to Al$_2$O$_3$ (right). In the former the dark layers correspond to alumina and the clear layers to the mixture of Al$_2$O$_3$-YTZP.

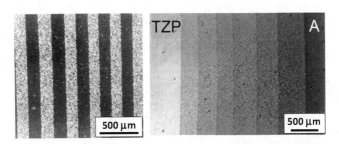

Figure 6. SEM microstructure of a multilayer of Al$_2$O$_3$/Al$_2$O$_3$-YTZP material (left) and a functionally graded material with discrete layers cchanging the compositon from YTZP (left) to Al$_2$O$_3$ (right). In the multilayer material, black layers are alumina and clear layers are Al$_2$O$_3$-YTZP.

The manufacture of planar substrates is commonly performed by the tape casting process, which consists in the preparation of a concentrated suspension that is spread over a surface covered by a flexible sheet that serves for rolling the full-length and avoids direct contact and sticking. As carrier films Mylar, tempered glass, etc, are employed. The blade is adjusted to provide the desired thickness of the tape. After casting, the solvent evaporates thus allowing the rolling and storage of the tape.[25-27] The cast tape must have the consistency and flexibility necessary to avoid cracking during handling. This is achieved with the addition of binders and plasticizers to typical concentrations of 10-20 vol.%. For this reason, most tape casting formulations are prepared in organic solvents, although water based systems have received increased interest in the last years. Typical thicknesses of the tapes are between 50-200 µm, although lower thicknesses can be also prepared. Typical formulations are prepared in organic solvents, being phosphate esters, glyceryl trioleate and menhaden fish oil the most popular deflocculants. Once a well-dispersed, stable slurry is prepared, the binding system (i.e., binders and plasticizers) is then added. Since competitive adsorption among the different additives may occur, the deflocculated suspension is first prepared usually by ball milling, and the binding system is added in a second milling step. Figure 7 shows schematically the sequence of steps followed in a tape casting process.

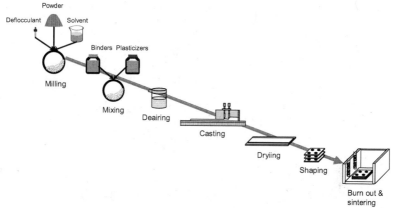

Figure 7. Typical sequence of steps in tape casting.

There is a big variety of formulations and components that have been produced by tape casting. Many applications lie in the field of electroceramics, including integrated circuitry, multilayer capacitors, isolating substrates, semiconductors, superconductors, solid oxide fuel cells, etc. A further advantage is that not only ceramic powders can be processed by tape casting, but also metals. As in any formulation, a first requirement is to prepare a well-dispersed, concentrated suspension. There are many papers dealing with tape casting of metals, and metal-ceramics, in particular Ni and Ni-Al$_2$O$_3$ and Ni-ZrO$_2$, for structural applications and for SOFCs, respectively. According to the flow chart of the process, shown in figure 7, the basic requirement is to have concentrated suspensions readily flowable. Figure 8 shows the flow curves of aqueous suspensions of submicronic Ni (average size of 2.5 μm, INCO T110, Canada) prepared to a solids content of 30 vol% before and after the addition of 5, 10, and 15 wt% of acrylic binders (a mixture 1:1 of Duramax B1000 and Duramax B1050, Rohm & Haas, USA).[28] The decrease of viscosity is related to the reduction of solids loading with increasing binder addition, as an extra amount of water is also added. In figure 9 the SEM microstructure of a monolithic specimen of Ni treated at 900°C/1h in Ar atmosphere is shown.

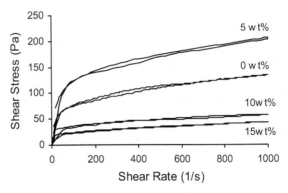

Figure 8. Flow curves of Ni aqueouss suspensions without and with 5, 10, and 15 wt% binders.

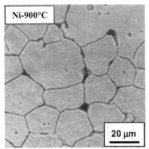

Figure 9. SEM microstructure of a monolithic specimen of Ni treated at 900°C/1h in Ar.

It is worthy to note that this suspensions containing binders strongly improve the wettability of metal substrates. As a consequence, these suspensions can be used as coatings on metal substrates. A possible application is the manufacture of permeable coatings in the inner walls of steel pipes, as those used as collectors in solar plants. The coating must be porous to allow migration of vapors to heat all the inner surface of the tubes when a hot flow of water is injected in the tube at high pressure. Figure 10 shows a collector steel tube coated with a Ni suspension by rolling. The microstructure of the steel/Ni interface after sintering at 650°C in CO_2 atmosphere is also shown demonstrating the good adherence and quality of the coating.[29]

Figure 10. Green coating of a steel tube with a tape casting suspension of Ni and microstructure of the steel/nickel interface. after sintering at 650°C in CO_2

More complex structures can be obtained mixing the different processes or the different slip formulations. The next example is the fabrication of porous Ni-Al_2O_3 coatings on Ni substrates. Figure 11 shows the SEM microstructures of porous coatings prepared by dipping a commercial Ni foil in suspensions of Ni-Al_2O_3 with 15 wt% Al_2O_3 prepared to different solids loadings (20, 25, and 30 wt%) and sintered at 1450°C/3h in N_2 flow.[30]

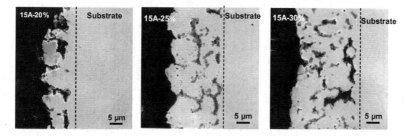

Figure 11. Ni foils coated by immersion in a suspension of Ni-Al_2O_3 with 15 wt% Al_2O_3 prepared to different solids loadings (20, 25, and 30 wt%) and sintered at 1450°C/3h in N_2 flow.

Similarly, Ni-YTZP materials have been prepared by colloidal routes in water. As an example, Ni-YTZP suspensions were prepared in water with a binder to promote wetting and adherence to the substrate. Suspensions contained pure Ni and mixtures with YTZP in increasing concentrations of 15, 30, and 50 vol% YTZP and a Ni foil was introduced in the different suspensions through a multistep dipping process. After sintering at 1200°C/1h a continuous FGM is formed, as shown in the microstructure of figure 12.[31]

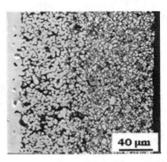

Figure 12. continuous FGM of Ni-YTZP (up to 50 vol% YTZP) produced by a multistep dipping in suspensions with 0, 15, 30, and 50 vol% YTZP.

In summary, it has been shown that colloidal processing routes are suuitable not only for ceramic processing but also for metals like Ni and metal-ceramic composites, such as Ni-Al$_2$O$_3$ and Ni-YTZP. Aqueous suspensions can be prepared at different solids loading depending on the shaping process to be used. Bulk bodies can be easily obtained by slip casting, as well as multilayers with total thicknesses of several milimeters. Laminates and FGMs with lower thickness (few decades of micrometers) are better produced combining tape casting formulations and dipping procedures. This allows to obtain laminates with different thickness and porosity or continuous FGMs by means of a multistep dipping process in suspensions with changing formulations.

ACKNOWLEDGEMENTS
This work has been supported by Spanish Ministry of Science and Innovation (MICINN, MAT2009-14369-C02-01.

REFERENCES
[1]M. Ashby, H. Shercliff, and D. Cebon, *Materials. Engineering, science, processing and design.* Elsevier, Oxford (UK), 2007.
[2]E. A. Mari, *Los Materiales Cerámicos*. Alsina, Buenos Aires (Argentina) 1998.
[3]R. W. Hertzberg, *Deformation and Fracture of Engineering Materials*, 3rd ed., Wiley, New York (USA), 1989.
[4]D. W. Richerson, "Advanced Ceramic Materials", pp. 65-88 in *Handbook of Advanced Materials*. Ed. by J. K. Wessel. John Wiley & Sons, Inc. Hoboken, NJ (USA), 2004.
[5]J. S. Reed. "*Introduction to the Principles of Ceramic Processing,*" 2nd ed, John Wiley & Sons, New York, 1995.
[6]J. S. Moya, S. Lopez-Esteban, and C. Pecharromán, The Challenge of Ceramic/Metal Microcomposites and Nanocomposites, *Prog. Mater. Sci.*, **52**, 1017-1090 (2007).
[7]W. D. Callister, *Materials Science and Engineering, An Introduction*, 7th ed., John Wiley, New York (USA) 2007.
[8]H. M. Chan, Layered Ceramics: Processing and Mechanical Behavior, *Annu. Rev. Mater. Sci.,* **27**, 249-282 (1997).
[9]P. Knauth and J. Schoonmann (editors), *Nanostructured Materials. Selected Synthesis Methods, Properties and Applications*. Kluwer Academic Press, New York (USA) 2004.
[10]F. Caruso, *Colloids and Colloid Assemblies: Synthesis, Modification, Organization and Utilization of Colloidal Particles*. Wiley-VCH (Germany) 2003.

[11]G. Y. Onoda and L. L. Hench (editors), *Ceramic Processing Before Firing*, John Wiley & Sons, New York (USA) 1978.

[12]M. N. Rahaman, *"Ceramic Processing and Sintering,"* 2nd ed, Marcel Dekker, New York (USA) 2003.

[13]G. Y. Onoda, "The Rheology of Organic Binder Solutions," pp. 235-251 in *Ceramic Processing Before Firing*. Edited by G. Y. Onoda and L. L. Hench, Wiley, New York, 1978.

[14]D. J. Shanefield, *"Organic Additives and Ceramic Processing,"* 2nd ed., Kluwer Academic Publishers, Boston, MA, 2000.

[15]R. Moreno, The Role of Slip Additives in Tape-Casting Technology: Part 1. Solvents and Dispersants, *Am. Ceram. Soc. Bull.*, **71** [10] 1521-31 (1992).

[16]R. Moreno, The Role of Slip Additives in Tape-Casting Technology: Part 2. Binders and Plasticizers, *Am. Ceram. Soc. Bull.*, **71** [11] 1647-57 (1992).

[17]R. J. Hunter, *Foundations of Colloid Science, Vol. 1*, Clarendon Press, Oxford (UK) 1987.

[18]J. N. Israelachvili, *Intermolecular and Surface Forces*. Academic Press, London (UK) 1985.

[19]D. J. Shaw *Introduction to Colloid and Surface Chemistry*, 4th edn, Butterworth-Heinemann, Oxford, Boston (USA) 1992.

[20]D. H. Everett, *Basic Principles of Colloid Science*. The Royal Society of Chemistry London (UK) 1988.

[21]O. Burgos-Montes, M. I. Nieto, and R. Moreno, Mullite Compacts Obtained by Colloidal Filtration of Alumina Powders Dispersed in Colloidal Silica Suspensions, *Ceramics Int.*, **33** [3] 327-332 (2007).

[22]R. Moreno, Tendencias en el Conformado a partir de Suspensiones, *Bol. Soc. Esp. Ceram. Vidr.*, **39** [5] 601-608 (2000).

[23]J. Requena, R. Moreno, and J. S. Moya, Alumina and Alumina/Zirconia Multilayer Composites Obtained by Slip Casting, *J. Am. Ceram. Soc.*, **72** (8) 1511-1513 (1989).

[24]J. S. Moya, A. J. Sanchez-Herencia, J. Requena, and R. Moreno, Functionally Gradient Ceramics by Sequential Slip Casting, *Mater. Lett.*, **14**, 333-35 (1992).

[25]R. E. Mistler and E. R. Twiname, *Tape Casting. Theory and Practice*, Am. Ceram. Soc. Inc., Westerville, OH, (USA) 2000.

[26]D. Hotza and P. Greil, Aqueous Tape Casting of Ceramic Powders, Mater. Sci. Eng. A, **202** [1-2] 206-217 (1995).

[27]T. Chartier and T. Rouxel, Tape-Cast Alumina-Zirconia Laminates: Processing and Mechanical Properties, *J. Eur. Ceram. Soc.*, **17** [2-3] 299-308 (1997).

[28]A. J. Sánchez-Herencia, C. A. Gutiérrez, A. J. Millán, M. I. Nieto, and R. Moreno, Colloidal Forming of Metal/Ceramic Composites, Key Eng. Mater., **206-213**, 227-230 (2002).

[29]B. Ferrari, J. L. Rodriguez, E. Rojas, and R. Moreno, Conformado por Vía Coloidal de Recubrimientos en la Cara Interna de un Tubo de Acero, *Bol. Soc. Esp. Ceram. Vidr.*, **43** (2) 501-05 (2004)

[30]I. Santacruz, B. Ferrari, M. I. Nieto, and R. Moreno, Ceramic Films Produced by a Gel-Dipping Process, *Adv. Eng. Mater.*, **5** (9) 647-650 (2003).

[31]I. Santacruz, B. Ferrari, M. I. Nieto, and R. Moreno, "Graded Ceramic Coatings Produced by Thermogelation of Polysaccharides, *Mater.Lett.*, **58** [21] 2579-82 (2004).

PROCESSING OF NITRIDE POROUS CERAMIC COMPOSITES VIA HYBRID PRECURSOR SYSTEM CHEMICAL VAPOR DEPOSITION (HYSYCVD)/DIRECT NITRIDATION (DN)

M. I. Pech-Canul and J. C. Flores-García

Centro de Investigación y de Estudios Avanzados del Instituto Politécnico Nacional "CINVESTAV-Zacatenco", Av. Instituto Politécnico Nacional 2508, Col. San Pedro Zacatenco, México, D.F., México, C.P. 07360
"CINVESTAV-Saltillo", Carr. Saltillo-Monterrey km 13, Saltillo, Coahuila, México, C.P. 25900

ABSTRACT

The synthesis of different kinds of composites via hybrid precursor system chemical vapor deposition/direct nitridation (HYSYCVD/DN) in a multi-step approach was investigated. Ceramic porous performs (SiC and Si_3N_4) were used as substrates and infiltrated in successive stages (S1-1, S1-2 and S2) varying processing conditions (type and flow rate of nitrogen atmosphere, temperature, and time). The composites were characterized by XRD and SEM. Also, some physical properties were determined by means of He picnometry, volume-measurement approach and immersion in Hg. Moreover, the thermodynamic feasibility of the possible reactions within the reaction systems was calculated using the FactSage® software and databases. The results show that the synthesis of nitride porous composites was influenced by test conditions at each stage of processing. Likewise, the effect of ceramic substrate type was important on the formation of nitrides and their microstructural characteristics (morphology, size and distribution), as well as on their physical properties.

INTRODUCTION

Porous ceramic composites are a class of materials not so uncommon. They are sought after for a great variety of technological applications, both at room and high temperatures. Typical applications include: molten metal filters, radiant burners, catalyst supports, dust collectors, heat exchangers, sensors, etc. [1, 2]. And from the broad spectrum of engineering materials available for such applications, silicon carbide (SiC) and silicon nitride (Si_3N_4) are frequently preferred. This is no surprise because both possess attractive thermomechanical properties, such as high strength at high temperatures, thermal shock and corrosion resistance. The latter is recognized in addition, because of its remarkable fracture toughness compared to other ceramics.

But perhaps what sounds unusual is the term "porous ceramic composite". It embraces three different concepts which can be defined individually. And from the point of view of materials science and engineering, it is expected that this type of materials must outperform any simple ceramic or porous material. An additional but not less important feature, in terms of structural applications, is the role that the constituents should play either as the matrix, or as the reinforcement.

Recently, there has been great interest in the development of porous ceramic composites containing SiC and Si_3N_4, and in consequence, a number of processing routes have been reported in the related literature. These comprise: carbothermal nitridation [2], extrusion process [3], gel casting [4], electroless deposition process [5], and chemical vapor infiltration (CVI) process [6]. Interestingly enough, in the case of Si_3N_4 composites, seeds of silicon nitride are used. Undoubtedly, the use of seeds as starting materials makes the process costly. Therefore, it is better to use processing routes that produce high purity silicon nitride in the absence of seeds of this ceramic. HYSYCVD/DN is an attractive route with the ability to produce silicon nitride and oxynitride without the need of seeds of these materials. It is based on the former HYSYCVD process recently reported [7], but in combination with a conventional direct nitridation process. It is clear that in order to exploit the HYSYCVD/DN process to its full potential, a study of the effect of the processing parameters on the type of phases formed, microstructure and resulting properties is definitely pertinent. Of great importance is to determine the effect of the substrate on the type of phases formed and other microstructure characteristics like size and morphology. And as regards to microstructure subject matter, the use of multiple steps in chemical vapor deposition is a simple and practical approach to study the microstructural evolution. This approach allows establishing important parameters such as the processing time to produce specific morphologies and/or sizes, including nanometer and micrometer scale phases. Also, because of the use of two different deposition mechanisms (HYSYCVD and DN), it is important to elucidate the influence of such mechanisms on the phases formed (Si_3N_4 and/or silicon oxynitride (Si_2N_2O)) and on the specific characteristics, like shape and size. In this particular investigation, the authors report on the processing, microstructural study and physical properties of $SiC/Si_2N_2O/Si_3N_4$ and Si_3N_4/Si_3N_4 nitride porous ceramic composites using the HYSYCVD/DN method, and obtained by a multiple infiltration approach in successive stages and in different nitrogen atmosphere types.

EXPERIMENTAL PROCEDURES

HYSYCVD/DN processing of composites

Ceramic cylindrical preforms (3 cm in diameter x 1.25 cm high) made out of Si_3N_4 (60 % porosity) or SiC (50 % porosity), and sodium hexafluorosilicate powders (Na_2SiF_6) cylindrical compacts were prepared by uniaxial compaction. The SiC powders were supplied by *Microabrasivos de México*, whereas the Si_3N_4 and Na_2SiF_6 by *Sigma Aldrich Inc*. Processing of composites was carried out in a hybrid CVD reactor, which consists of a horizontal tube furnace with end-cap fittings, gas inlets and outlets to supply the nitrogen precursor as well as with devices to control flow rate, pressure and process atmosphere. The ceramic porous performs (Si_3N_4 and SiC) were positioned within the alumina tube at the high temperature zone (at the center of the reaction chamber), while the Na_2SiF_6 compacts were placed in the low temperature zone nearby the gas entrance (see Figure 1).

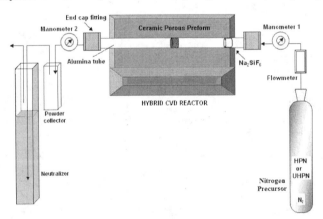

Figure 1. Schematic representation of the experimental set-up.

For the processing of composites, ceramic porous preforms (in Si_3N_{4p} and SiC_p, the subscript p stands for particulate) were infiltrated in subsequent stages (S1-1, S1-2 and S2).

Stage 1 (S1- synthesis and growth of new phases)

This stage was performed in 2 sub-stages (S1-1 and S1-2) by alternating the preform deposition sides from one to another stage. The initial stage (S1-1) and the subsequent stage S1-2 were performed in high purity nitrogen (HPN). The ceramic porous preforms were heated in the nitrogen atmosphere at a rate of 15 °C/min up to 1300 °C and then maintained at this temperature for 70 min at a constant gage pressure of 13 mbars. Simultaneously, Na_2SiF_6 was heated (in the temperature range 250-550 °C) and decomposed to generate the silicon-fluorine (SiF_X, x=0-4) gaseous reacting species necessary for the formation of Si_3N_4 and/or Si_2N_2O, and their subsequent deposition over the ceramic substrates.

Stage 2 (S2- further growth and consolidation of nitrides formed)

In this stage, the samples processed in the first stage (S1-1 and S1-2), were treated under the following conditions: heating rate of 15 °C/min, maintained isothermally at 1350 °C for 120 min in ultra high purity nitrogen (UHPN) atmosphere at a flow of 30 cm^3/min, at a constant pressure in the system of 13 mbars. It is noteworthy, that at this stage Na_2SiF_6 was not used, thus consisting of a direct nitridation stage. It should be noted that the processing conditions used in the current work were optimized in previous investigations, using only one infiltration step [8].

Porous composites physical properties evaluation

The apparent density (ρa) and apparent volume (Va) of porous samples were determined by helium Pycnometry using a Quantachrome Instruments model Multipycnometer semi-automatic helium (He) pycnometer. The test was conducted at a constant He pressure calibration of 0.117 MPa, with high purity He (99.999 %) supplied to the system. For infiltration percentage (infiltration %) and residual porosity (ε_r %) determination, apparent volume data obtained from measurements of helium pycnometer were used. The total volume (Vt) of the composites was calculated based on their geometry by measurement of its dimensions with a Mitutoyo digital Vernier calliper. Bulk density (ρ_b) was calculated by means of volume-measurement approach and by immersion in Hg using Archimedes's principle.

RESULTS AND DISCUSSION

Specimen phase analysis and microstructure

X-ray diffraction patterns of specimens before and after the successive stages of processing are shown in figures 2 (SiC) and 3 (Si_3N_4). In figure 2 it can be seen that before the treatment reflections pertaining only to SiC (JCPDS-74-1302) particles are revealed; after the processing, the presence of silicon nitride and oxynitride reflections - besides SiC - is also observed. According to the analysis by XRD of processed specimens, Si_2N_2O (JCPDS-83-2142), α–Si_3N_4 (JCPDS-76-1409) and β–Si_3N_4 (JCPDS-71-0623) phases were formed on the SiC_P porous preforms. With respect to Si_3N_4 specimens, in figure 3 it can be seen that in the X-ray diffractograms there was no change in the type of phases present initially (α and β–Si_3N_4) with respect to the samples treated, but a variation in the intensity of reflections is noted, indicating that new amounts of α and β–Si_3N_4 phases were formed.

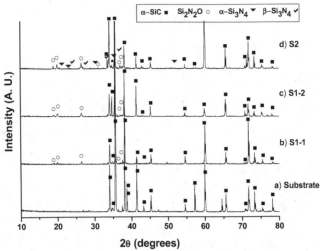

*Figure 2. X-ray diffraction patterns of SiC specimens before and after successive stages of processing:
a) before processing, b) stage S1-1 in HPN, c) stage S1-2 in HPN, and d) stage S2 in UHPN.*

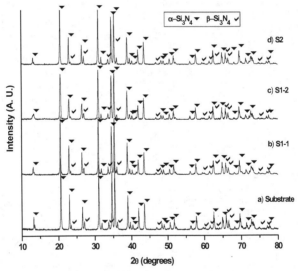

*Figure 3. X-ray diffraction patterns of Si_3N_4 specimens before and after successive stages of
processing: a) before processing, b) stage S1-1 in HPN, c) stage S1-2 in HPN, and d) stage S2 in
UHPN.*

These results prove to that unlike SiC, Si_3N_4 showed a high resistance to be oxidized, since Si_2N_2O is absent. In this regard the contribution of direct nitridation mechanism in the formation of new phases can be considered minimum; this implies that silicon nitride substrate does not react with nitrogen during processing.

Results from phase semi-quantitative analysis (using the X-ray diffractograms) of both substrates processed are shown in figures 4 and 5. The chart in figure 4 shows an increase in the amount of nitrides formed during the successive infiltration stages on SiC substrates. From these results, it can be observed that in the initial stage (S1-1) small amounts of Si_2N_2O were generated, which act as nucleation sites on the SiC substrates for further growth and new phases; that is confirmed in the subsequent stage S1-2, with an increase of Si_2N_2O of about twice the initial amount. Finally in S2 stage, it is observed a continuous Si_2N_2O growth, accompanied by α-Si_3N_4 y β-Si_3N_4 formation. This can be due to a decrease in the oxygen partial pressure into reaction chamber and by oxygen desorption (see eq. 4 and 5).

Regarding Si_3N_4 substrates, the chart in figure 5 shows a variation in α and β-Si_3N_4 quantities present at each processing stage. At this point it is pertinent to mention that the α-Si_3N_4 $\rightarrow$ β-Si_3N_4 polymorphic transformation is irreversible [9], indicating that an increase of α-Si_3N_4 phase at the expense of a decrease of β-Si_3N_4 present initially in the substrate is not likely. In the processing of silicon nitride at high temperature, it is common that β-Si_3N_4 phase increases due to the aforementioned transformation, therefore, these results are strong evidence that a new phase is deposited (in this case α-Si_3N_4) on silicon nitride substrates, showing the effectiveness of processing by HYSYCVD.

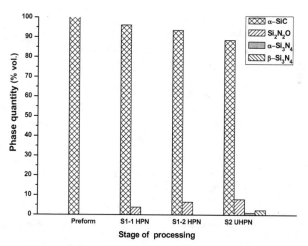

Figure 4. Phase quantities into SiC specimens processed in the successive stages of infiltration (S1-1 $\rightarrow$ S1-2 $\rightarrow$ S2) in HPN and UHPN.

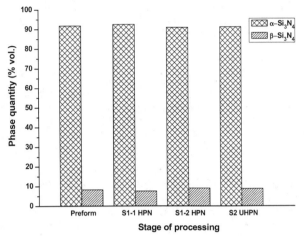

Figure 5. Phase quantities into Si_3N_4 specimens processed in the successive stages of infiltration (S1-1 → S1-2 → S2) in HPN and UHPN.

The thermodynamic feasibility of possible nitridation reactions occurring in the different systems is showed in equations (1)-(6).

In SiC_p substrates, the formation of silicon oxynitride, α– and β-Si_3N_4 generated can be attributed to:

1) SiC substrate direct oxynitridation and nitridation.

$$4SiC_{(s)} - 2N_{2(g)} + 3O_{2(g)} - 2Si_2N_2O_{(s)} + 4CO_{(g)} \quad (1)$$
$$\Delta G°_{1300°C} = -1,644 \text{ kJ/mol}$$

$$5SiC_{(s)} + 3N_{2(g)} + 3O_{2(g)} - Si_2N_2O_{(s)} + Si_3N_{4(s)} + 5CO_{(g)} \quad (2)$$
$$\Delta G°_{1300°C} = -1,621 \text{ kJ/mol}$$

2) The reaction of Si-F_x species (generated by the decomposition of Na_2SiF_6) with the N_2 and O_2 present in the system. According to recent publications [8, 10], oxygen content in the nitrogen precursor [HPN (<10 ppm) and UHPN (<3.5 ppm) of O_2 traces] [11] is sufficient for inducing the formation of silicon oxynitride in gas phase.

$$6SiF_{4(g)} + 6SiF_{3(g)} - 6SiF_{2(g)} + 6SiF_{(g)} - 6Si_{(g)} - 18N_{2(g)} + 3O_{2(g)} - 6Si_2N_2O_{(s)} + Si_3N_{4(s)} - nNCP_{(g)} (3)$$
$$\Delta G°_{1300°C} = -5,176 \text{ kJ/mol}$$

* nNCP stands for moles of non-condensed phases (gaseous species formed or without reacting)

As the oxygen in the reaction chamber decreases with time, the amount of silicon nitride formed with respect to silicon oxynitride increases.

$$6SiF_{4(g)} - 6SiF_{3(g)} - 6SiF_{2(g)} + 6SiF_{(g)} - 6Si_{(g)} - 18N_{2(g)} + 1.5O_{2(g)} - 3Si_2N_2O_{(s)} + 3Si_3N_{4(s)} - nNCP_{(g)} (4)$$
$$\Delta G°_{1300°C} = -4,339 \text{ kJ/mol}$$

3) Nitridation of Si_2N_2O phases already formed through a conversion from oxynitride by oxygen desorption.

$$6Si_2N_2O_{(s)} + 2N_{2(g)} - Si_3N_{4(s)} + nNCP_{(g)} \quad (5)$$
$$\Delta G°_{1350°C} = -37.43 \text{ kJ/mol}$$

With respect to Si_3N_4 substrates, formation of different quantities of $\alpha-$ and β-Si_3N_4 during the processing in all the stages was mainly due to gas-gas and solid-gas reactions of Si-F_x species with the N_2 atmosphere and Si_3N_4 within the system.

$$6SiF_{4(g)} - 6SiF_{3(g)} - 6SiF_{2(g)} + 6SiF_{(g)} + 6Si_{(g)} + 18N_{2(g)} - 5Si_3N_{4(s)} + nNCP_{(g)} \quad (6)$$
$$\Delta G°_{1300°C} = -3,527 \text{ kJ/mol}$$

In this sense, a certain amount of α-Si_3N_4 was initially deposited via HYSYCVD, acting as nucleation sites for further growth, which is associated with the $\alpha \rightarrow \beta$-Si_3N_4 polymorphic transformation at high temperature in nitrogen atmosphere.

Regarding the microstructural characterization, figures 6 to 9 show the results of SEM analysis of SiC and Si_3N_4 specimens before and after the processing in successive stages of infiltration.

Figure 6a shows the photomicrographs corresponding to the SiC substrates, which have an irregular geometry and a wide particle size distribution (47 μm in average), while figure 6b shows Si_2N_2O deposits formed on the SiC particles in S1-1 stage; these deposits show a sponge-like coating morphology followed by needle-like growth of pin-like short fibers with a maximum diameter of 100 nm and lengths up to 5 μm.

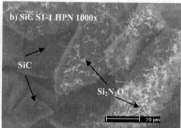

Figure 6. SEM photomicrographs of SiC specimens: a) before processing and b) after processing in stage S1-1 in HPN.

In figure 7a the photomicrograph of specimens processed in stage S1-2 shows an increase in the amount of deposited Si_2N_2O; the dimensions of pin-like fibers reached diameters of 500 nm and lengths up to 20 μm. A change of appearance from spongy to compact coatings is also noticed. Finally, in stage S2 (figure 7b) the amount of nitrides increases significantly, observing α and β-Si_3N_4 nanowires randomly distributed around the composite.

Figure 7. SEM photomicrographs of SiC after successive stage of processing: a) S1-2 in HPN and b) S2 in UHPN.

With respect to Si_3N_4 microstructural characterization, figures 8 and 9 show that in all processing stages, deposits of α and β–Si_3N_4 were formed onto the substrate particles.

Figure 8a shows the photomicrograph corresponding to the Si_3N_4 substrates, which have an irregular geometry and ~1μm average particle size, while figure 8b shows that new Si_3N_4 deposits formed (mainly α–Si_3N_4) on substrate particles in S1-1 have fine-grained polycrystals morphology, generating compact coatings.

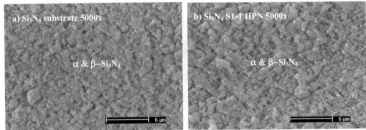

Figure 8. SEM photomicrographs of Si_3N_4 specimens: a) before processing and b) after processing in stage S1-1in HPN.

In figure 9a the photomicrograph of specimens processed in S1-2 shows that more silicon nitride phases were deposited onto compact coatings formed in S1-1 causing the growing of the grained polycrystals. Finally, figure 9b shows that the deposition of polycrystals caused the formation of particulate agglomerates giving rise to a consolidate aspect.

Figure 9. SEM photomicrographs of Si₃N₄ after successive stage of processing: a) S1-2 in HPN and b) S2 in UHPN.

The variation in morphology, size and distribution of phases present in the SiC and Si₃N₄ substrates processed at various processing stages and under different nitrogen atmospheres, as shown in the SEM analysis, can be attributed to both temperature and supersaturation condition (concentration of atoms or molecules adsorbed on the substrate surface) [12].

Physical properties

Results from physical properties determination are showed in Table 1. These results show that the bulk density values measured by different methods have similar behavior. By comparing the amount of phases formed on the different types of substrates, it can be noted that the highest infiltration percentage (infiltration %) occurs in Si₃N₄ composites compared with those of SiC, indicating that the deposit of new phases of nitrides by HYSYCVD is more effective on this type of substrates (silicon nitride). The SiC/Si₂N₂O/Si₃N₄ and Si₃N₄/Si₃N₄ composites exhibit ~ 46 % and 50 % of residual porosity, respectively.

Table 1. Physical properties of SiC/Si₃N₄/Si₂N₂O and Si₃N₄/ Si₃N₄ porous composites.

Sample	Apparent density ρa (g/cm³)	Bulk density ρ_b (g/cm³)	Residual porosity εr (%)	Infiltration (%)	Hg bulk density ρ_{bHg} (g/cm³)
SiC/Si₂N₂O/Si₃N₄	3.07	1.63	46.85	3.14	1.52
Si₃N₄/Si₃N₄	2.82	1.38	50.72	9.27	1.37

SUMMARY AND CONCLUSIONS

In this work, results from the processing, microstructural characteristics, and physical properties of SiC/Si₂N₂O/Si₃N₄ and Si₃N₄/Si₃N₄ porous ceramic composites, processed in a multi-step approach and different nitrogen atmospheres – via hybrid precursor system chemical vapor deposition (HYSYCVD) and direct nitridation (DN) – are shown. The XRD characterization reveals formation of Si₂N₂O, α-Si₃N₄ and β-Si₃N₄ phases on SiC specimens, whereas in Si₃N₄ specimens experimental conditions used show the formation of varying amounts of α- and β-Si₃N₄. In addition, according to the analysis by SEM, the Si₃N₄ and Si₂N₂O phases were deposited into SiC porous preforms in a variety of morphologies, ranging from sponge-like coatings and pin-like fibers to compact deposits and nanowires randomly distributed. Regarding to Si_3N_4 specimens, the analysis by SEM showed that the formation of α- y β-Si_3N_4 have a fine-grained polycrystals morphology, generating compact coatings. The $SiC/Si_2N_2O/Si_3N_4$ and Si_3N_4/Si_3N_4 composites exhibit $\sim$ 46 % and 50 % of residual porosity, respectively. The effect of ceramic substrate type was important on the formation of nitrides and their microstructural characteristics, as well as on their physical properties.

ACKNOWLEDGEMENTS

Authors gratefully acknowledge CONACyT for financial support under contract No. SEP-CONACYT-2005-1/24322. Also, Mr. J. C. Flores-García expresses his gratitude to CONACyT for providing a scholarship. Finally, the authors also thank Mrs. M. Rivas-Aguilar and S. Rodriguez-Arias for technical assistance during the analysis by SEM and XRD, respectively.

REFERENCES

[1] J-F. Yang, G-J. Zhang, N. Kondo and T. Ohji, "Synthesis and properties of porous Si_3N_4/SiC nanocomposites by carbothermal reaction between Si_3N_4 and carbon", *Acta Materialia*, 50 (2002), p. 4831-4840.

[2] J-F. Yang, S-Y. Shan, R. Janssen, G. Schneider, T. Ohji and S. Kanzaki, "Synthesis of fibrous β-Si_3N_4 structured porous ceramics using carbothermal nitridation of silica", *Acta Materialia*, 53 (2005), p. 2981-2990.

[3] A. K. Gain, J-K. Han, H-D. Jang and B-T. Lee, "Fabrication of continuously porous SiC-Si_3N_4 composite using SiC powder by extrusion process", *Journal of the European Ceramic Society*, 26 (2006), p. 2467-2473.

[4] W. Zhang, H. Wang and Z. Jin, "Gel casting and properties of porous silicon carbide/silicon nitride composite ceramics", *Materials Letters*, 59 (2005), p. 250-256.

[5] R. K. Paul, H-D. Jang and B-T. Lee, "Fabrication of platinum coating on continuous porous SiC-Si_3N_4 composites by the electroless deposition process", *Journal of Materials Processing Technology*, 209 (2009), p. 2958-2962.

[6] Y. Liu, L. Cheng, L. Zhang, Y. Xu and Y. Liu, "Design, preparation, and structure of particle performs for $Si_3N_{4(p)}$/Si_3N_4 radome composites prepared using chemical vapor infiltration process", *Materials*, 15 (2008), p. 62-66.

[7] A. L. Leal-Cruz, M. I. Pech-Canul and J. L. de la Peña, "A low-temperature and seedless method for producing hydrogen-free Si_3N_4", *Revista Mexicana de Física*, 54 (2008), p. 200-207.

[8] A. L. Leal-Cruz, "Thermodynamics, kinetics and microstructural study of Na_2SiF_6 decomposition-silicon nitrides formation in systems Na_2SiF_6-nitrogen precursor-diluent", Ph. D. Thesis, Cinvestav Saltillo, Saltillo Coah. México, (2007).

[9] F. L. Riley, "Silicon Nitride and Related Materials", *Journal of the American Ceramic Society*, 83 (2000) p. 245-265.

[10] M. I. Pech-Canul, J. L. de la Peña, and A. L. Leal-Cruz, "Effect of processing parameters on the deposition rate of Si_3N_4/Si_2N_2O by chemical vapor infiltration and the in situ thermal decomposition of Na_2SiF_6", *Appl. Phys. A.*, 89 (2007), p. 729-735.

[11] INFRA Group-gas supplier, "Product catalog", Mexico (2008).

[12] J. M. Blocher Jr., "Structure/property/process relationships in chemical vapor deposition CVD", *J. Vac. Sci. Technol*, 11 (1974), p. 680-686.

EFFECT OF NANO-SIO$_2$ ON MICROSTRUCTURE, INTERFACE AND MECHANICAL PROPERTIES OF WHISKER-REINFORCED CEMENT COMPOSITES

Mingli Cao and Jianqiang Wei
School of Civil Engineering, Dalian University of Technology
Dalian 116024, Liaoning, China

ABSTRACT

As a kind of microfiber, CaCO$_3$ whisker can improve the properties of cement efficiently. However, the weak interfacial bond and dispersity of whiskers in cement matrix severely restrained the further improvement of properties. Nanometer silica was introduced to modify and achieve interfacial optimization between whiskers and cement matrix to ensure the reinforcement of whiskers be given full play. Crystal structures, microcosmic appearances and characterizations of nano-silica, CaCO$_3$ whisker and the composite were studied by X-ray diffraction (XRD), scan electron microscope (SEM/EDS), and transmission electron microscope(TEM), respectively. Effect of different silica content on the microstructure, interface and mechanical properties of whisker-reinforced cement was investigated. The results show that the incorporation of nano-silica resulted in an increase in interfacial properties between whisker and cement matrix, the mechanical properties of composites and microstructure of the matrix were all improved. Furthermore, mechanisms of whsiker-cement's interface optimized by nano-silica was also discussed.

INTRODUCTION

As with ceramic, cement composites are also the most important inorganic materials, which have been widely applied in many fields such as: housing construction, road and bridge engineering, energy exploitation and national defense construction. Against their advantages of high strength, good integrality and low cost, however, the biggest defect of cement paste is the little toughness of hardened cement paste structure because this material is both fissile and brittle compared with other building materials. In order to improve the tensile strength, it is necessary to decrease the amount of microcracks and control its growth effectively[1], and the most effective way is to be reinforced by fibers. CaCO$_3$ whisker is a new kind of microfiber, with high strength, high modulus, high durability, low price, as well as environmental protection, which is an ideal substitute for the traditional fibers.

Whiskers have been reported to improve the properties of brittle materials. The related research was reported by Lan Sun and Jinsheng Pan, who obtained 120% and 63% increase in the flexural strength and fracture toughness respectively using TiC whiskers to reinforce MoSi matrix composites[2]. Zan Qingfeng, Dong Limin et al. reported the use of SiC whiskers in Al$_2$O$_3$/Ti$_3$SiC$_2$ multilayer ceramics, and showed that the gain in of strength and work of fracture for materials were about 4.4 and 19.6%, respectively[3]. Frank A. Müller, Uwe Gbureck, Toshihiro Kasuga et al. reported the reinforcement of calcium phosphate cements with hydroxyapatite whiskers, and obtained a rather perfect result that the flexural strength and work of fracture of cements increased by 60%to 7.4MPa and by 122% to 102 J/m^2, respectively[4].And in previous studies, we have reported that the the microstructure of Portland cement composites can be improved by $CaCO_3$ whiskers, and the flexural strength, impact strength and split tensile strength were increased by 39.7%, 39.25% and 36.34% at maximum, respectively. However, interfacial property, which limited the reinforcing effects of $CaCO_3$ whiskers in cement matrix, is the problem must be resolved. More energy will be consumed, and the mechanical properties will be further improved, if interfacial properties of whisker in cement matrix were increased.

Here, we initially report on the effect of nano-SiO_2 on the reinforcing effects of $CaCO_3$ whiskers in cement matrix, such as microstructure, interface and mechanical properties.

EXPERIMENTAL
Raw Materials
    $CaCO_3$ whiskers, nano-silica and 42.5# ordinary Portland cement were used in this study.
    The cement used in this study was produced by Dalian cement factory, with a specific surface area of 348.1 m^2 / kg .The $CaCO_3$ whiskers were synthesised by a simple carbonation method using calcium chloride as a precursor. Briefly, the calcium chloride was prepared by mixing magnesium chloride solution with certain concentration and calcium hydroxide prepared by lime slaking. 0.09~0.12 liters per minute carbon dioxide gas was blowed into the system to initiate carbonization reaction. Carbonation treatment was performed at 85±2.5°C for 5~6 h under the alkalinity environment(PH>9). The chemical constituents of $CaCO_3$ whiskers and the ordinary Portland cement were assayed and shown in table 1.

Table 1. Chemical constituent of $CaCO_3$ whiskers and Portland cement (wt.%).

composition	CaO	CO₂	MgO	SO₃	SiO₂	Al₂O₃	Fe₂O₃	K₂O	Na₂O	TiO₂	P₂O₅	MnO	SrO	Cr₂O₃
whiskers	54.93	42.07	2.14	0.31	0.29	0.11	0.07	—	—	—	—	—	0.05	0.03
nano-silica	0.81	—	0.95	0.84	93.47	0.16	0.10	2.89	0.23	—	0.40	0.06	—	—
cement	52.04	6.00	5.15	4.66	23.08	4.95	2.64	0.81	0.14	0.31	0.07	0.06	0.09	—

The macroscopic feature of $CaCO_3$ whiskers is white fluffy powder. In order to have a direct and vivid understanding of $CaCO_3$ whiskers, the microcosmic appearance and crystal structure of $CaCO_3$ whiskers were analyzed. Fig.1 show the microstructure and XRD pattern of $CaCO_3$ whiskers used to reinforce the Portland cement. It can been seen that, the whiskers were needle-like with high integrity. The whiskers exhibit an average length of 25μm at an aspect ratio of 30~50, and the most of fundamental peaks of XRD pattern can be indexed to the aragonite $CaCO_3$, from which it can be indicated that aragonite is the crystal form of $CaCO_3$ whisker with high purity and purity. Fig.2 shows the the microstructure of nano-silica used to improve interfacial properties between whisker and cement matrix. From the picture we can see that the particle diameters of silica fume ranges from 200 nm to about 10 nm.

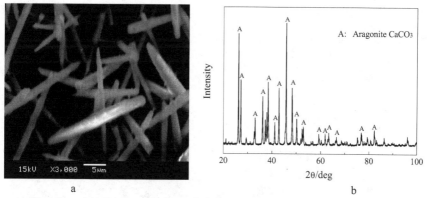

a   b

Fig.1 Scanning electron microscopy images and XRD pattern of CaCO$_3$ whiskers.

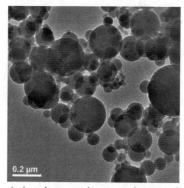

Fig.2 transmission electron microscopy images of nano-silica.

Specimen Preparation and Testing

Cement specimens for all experiments were prepared by mixed with single CaCO$_3$ whisker content, and different content of slag micropowder such as 0wt%(S-0), 2.0wt% (S-2), 4.0wt% (S-4), 6.0wt% (S-6), 8.0wt% (S-8) and 10.0wt% (S-10). Then the powder/whisker/cement and liquid systems were equably churned by cement paste mixer.

Then, the composites were agitated by cement paste mixer with 0.3 water/cement ratio by mass (W/C). The cement was cast into cubic moulds with dimensions of 40mm×40mm×40mm for compression and equivalent splitting tensile test. And the specimens with sizes of 40mm×40mm×160mm, respectively, were cast, being used for flexural test. The normal specimens were stored for 24 hours with the mold in standard curing box of cement. After demolding, the abovementioned specimens were stored under 100% humidity and 18 ℃ for 28 days.

The compressive strength of the specimens was measured by pressure machine. The equivalent splitting tensile test and flexural test were carried out in a computer controlled electro-hydraulic servo universal tester (WDW-50) with a crosshead speed of 0.05 mm/min according to Fig. 3 (Chinese Standard GBJ81-85).The impact test was carried out by a pendulum impact testing machine, and the microstructure of whiskers and fracture surfaces was characterized by scanning electron microscopy (SEM) (JSM-5600LV).

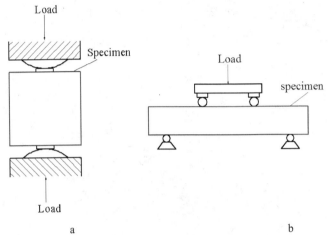

a   b

Fig.3. Sketch map of tests: (a) Equivalent splitting tensile test,(b) flexural test.

RESULTS AND DISCUSSION
Mechanical Properties
Split Tensile Strength
    The equivalent splitting tensile test has been done in this study, to research the tensile strength of cement feasibly. The value of split tensile strength was calculated according to Eq. (2).

$$T = 2F/\pi A \qquad (2)$$

Thereinto, T is split tensile strength (MPa), F is failure load (N), A is the area of cleavage plane.Each group includes three specimens. The arithmetic average value of the records shall as the measured value. If there is a value with relative deviations more than 15% of the median, the measured value should be the median.

    Table2 summarizes the split tensile strength of whisker-reinforced cement composites reinforced with various amounts of nano-silica, from which, we can see that incorporation of nano-silica resulted in an increase in the split tensile strength by 121.14% to 3.87Mpa for S-8. A further increase of the filler content led to a strength decrease to 3.79 MPa for S-10.When nano-silica content increases from 0% to 10.0%, there was also an increase in split tensile strength. It was found that the dispersing of fiber in cement matrix was also improved with the incorporation of nano-silica. Furthermore, the number of interface defect between whiskers and cement matrix decreased with the increase of nano-silica content, which may also account for the increase of split tensile strength.

Table 2. Effect of nano-silica content on the split tensile strength of whisker-reinforced cement

Sample	Specification	Split tensile strength (MPa)
S-0	Whisker- cement without nano-silica	1.75
S-2	+2.0 wt% nano-silica	2.11
S-4	+4.0 wt% nano-silica	2.72
S-6	+6.0 wt% nano-silica	3.42
S-8	+8.0 wt% nano-silica	3.87
S-10	+10.0 wt% nano-silica	3.79

Flexural Strength

Thus, four-point bending tests were performed to evaluate the influence of nano-silica fume content on the flexural properties of whisker-reinforce cement. The value of flexural strength was calculated according to Eq. (1).

$$R = FL/bh^2 \qquad (1)$$

Thereinto, R is flexural strength (MPa), F is failure load (N), L is span between two supports (mm), b is section width of specimen (mm) and h is section height of specimen (mm)[5]. Each group included three specimens. Values with relative deviations more than 15% were deleted from the valuable data, remaining data was used to calculate the average of the group. The effect of nano-silica on the flexural reinforcement of whisker reinforced cement composites was shown in Fig.4. The results shown in Fig.4 summarize that incorporation of slag micropowder resulted in an increase in the flexural strength, and it was increased gradually with the increased content of nano-silica slag micropowder. The maximal flexural strength increases by 55.929% for S-10.

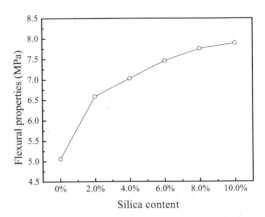

Fig.4. The effect of nano-silica on the flexural reinforcement of whisker reinforced cement composites

Interfacial Microstructures

To further investigate the interfacial-toughening mechanism by silica fume, the microstructure of whiskers drawn from matrix is observed. Fig. 5 shows the micro-appearance of CaCO$_3$ whiskers used to reinforce the Portland cement matrix and whisker-reinforced cement fractured surfaces with and without nano-silica. Whisker-reinforced cement fractured surface without slag nano-silica (S-0) was shown in Fig.5-a and Fig.5-b, whiskers as well as holes resulting from whiskers that were pulled out are homogeneously distributed throughout the cement matrix with smooth surface and without any residue, so we can arrive at a conclusion that in the composite, whisker pull-out is dominant, which indicates that the interfacial bond strength of whiskers/cement is not high enough. And this is exactly the reason why we research the improvement of nano-SiO$_2$ on the microstructure, interface and mechanical properties of whisker-reinforced cement composites

Fig.5-c and Fig.5-d show SEM images of fractured surfaces of S-4 and PW-8. Fig.4-c shows a whisker that was oriented perpendicular to the crack plane and was pull-out of the matrix..Be different from the wiskers pull out from S-0, the whisker was pulled out of the cement matrix with some residual cement particles bonded to the whisker in S-4, and there were more residual cement particles with the increasing of silica content. From the interface of whisker pulled out form cement in S-8, we can see that the whiskers were all tightly wrapped by a thick layer of cement. During the pull out, energy that would normally cause crack propagation is partially expended by debonding and by friction, which have been improved, as the whisker slides against adjacent microstructure features[6]. The matrix adhering to the fiber can be worn during pullout process and may cumulate near fiber end. The cumulative matrix remnant in turn contributes to the resistance to fiber pullout load[7]. Therefore, the greatly enhanced pullout energy is believed to be a reflection of this mechanism.

Fig.5 The effect of nano-silica on the interfacial microstructures of whisker reinforced cement composites

And from the Fig.6-a and Fig.6-b, which show the energy dispersive spectroscopy (EDS) analysis of the interface between CaCO3 whiskers and cement matrix of S-0 and S-8, respectively, we can see that the major element of the surfaces are calcium, oxygen, carbon and silicon, but the the biggest difference between the the is that the proportion of calcium and silicon. In the surface of S-0, carbon dominating silicon, however, in S-8 the reverse seems to be happening, which indicates that the bond strength was high enough to peel off part of cement from the matrix. From all the SEM and EDS results and analysis, we can conclude that the microstructure and the interfacial bond strength were effectively improved with nano-silica, which can effectively increase the work of fracture/fracture toughness.

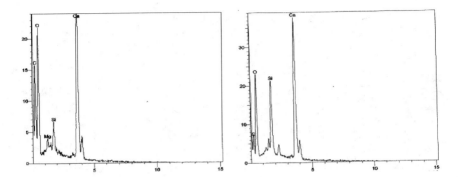

Fig.6. The energy dispersive spectroscopy (EDS) analysis of the interface between CaCO3 whiskers and cement matrix.

Interfaces between fibers and cement matrix are the weakest parts in fiber-reinforced cement composites, which limited the reinforcement of fiber to cement. The pozzolanic effect of nano-silica fume can improved the microstructure of interfaces and width of the interfacial transition zone, by producing more densified and homogenous C-S-H in in the process of cement hydration. The silica fume improves pore structure in two ways: its small particle size results in a filler effect in which the silica fume particles bridge the spaces between cement grains and the spaces between cement grains and aggregate; and the silica fume reacts pozzolanically with calcium hydroxide to produce a greater solid volume of C-S-H gel, leading to an additional reduction in capillary porosity during hydration[8]. It is for these chemical and physical reasons that the reinforcement of wiskers in cement matrix was improved effectively.

CONCLUSION

1. Comparing to that of the matrix without silica fume, the enhancement in pullout energy due to silica fume is more significant than that in toughness. Incorporation of nano-silica fume resulted in an increase in the mechanical properties of whisker-reinforced cement and it was increased gradually with the increased content of nano-silica fume. The incorporation of nano-silica resulted in an increase in the split tensile strength by 121.14% to 3.87Mpa, and the maximal flexural strength was increased by approximately 56.0%.

2.The microstructure of wiskers pulled out from cement matrix reveals a great amount of cementitious materials adhering to whisker surface. Consequently, the cementitious material contributes to the friction and resistance during the whiskers pullout process. The microstructure and interfacial bond strength of whiskers/cement were improved efficiently, therefore more energy was expended when a certain amount of nano-silica fume content is incorporated.

3. The reinforcement of wiskers in cement matrix was improved effectively for the chemical and physical reasons of nano-silica fume, such as the pozzolanic effect and micro-aggregate filling effect.

REFERENCES

[1]Yao Wu, Zhong Wenhui. Effect of polypropylene fibers on the long-term tensile strength of concrete[J]. Journal of Wuhan University of Technology--Materials Science Edition, 2007, 22(1), 52-55

[2] Lan Sun, Jinsheng Pan. TiC whisker-reinforced MoSi$_2$ matrix composites[J]. Materials Letters, 2001, 51(3): 270–274.

[3]Zan Qingfeng, Dong Limin, Wang Chen, et al. Improvement of mechanical properties of Al$_2$O$_3$/Ti$_3$SiC$_2$ multilayer ceramics by adding SiC whiskers into Al$_2$O$_3$ layers[J]. Ceramics International, 2007, 33(3): 385-388

[4]Frank A. Müller, Uwe Gbureck, Toshihiro Kasuga, etc. Whisker-Reinforced Calcium Phosphate Cements[J]. Communications of the American Ceramic Society , 2007, 90(11): 3694-3697

[5]Yiping Ma, Beirong Zhu, Muhua Tan. Properties of ceramic fiber reinforced cement composites[J]. Cement and Concrete Research, 2005, 35(2): 296– 300

[6]D. W. Richerson. "Modern Ceramic Engineering," New York : Marcel Dekker, 1992, pp.731–807.

[7]Yin-Wen Chan and Shu-Hsien Chu, Effect of silica fume on steel fiber bond characteristics in reactive powder concrete, Cement and Concrete Research, Volume 34, Issue 7, July 2004, Pages 1167-1172

[8]D.P. Bentz and P.E. Stutzman, Evolution of porosity and calcium hydroxide in laboratory concretes containing silica fume. *Cem. Concr. Res.* 24 (1994), pp. 1044–1050. Abstract | View Record in Scopus | Cited By in Scopus (18)

Directional Solidification and Microwave Processing

ENGINEERED SELF-ORGANIZED MICROSTRUCTURES USING DIRECTIONAL SOLIDIFICATION OF EUTECTICS

V.M. ORERA, J.I PEÑA, A. LARREA, R.I.MERINO, P.B. OLIETE
Institute of Materials Science in Aragón, Fac. Of Sciences, University of Zaragoza
Zaragoza, Spain

ABSTRACT

Using Directional Solidification of Eutectic Ceramics (DSEC) different morphologies and microstructures of the eutectic composites ranging from well known fibrous and lamellar regular patterns to 3d-entangled interpenetrating or even the unusual split-ring resonator-like morphologies can be fabricated using several growth-from-melt techniques. Changing the eutectic components and/or the microstructure morphology we can develop self-organised materials exhibiting new properties. We describe here the fabrication of some binary and ternary DSEC showing for example new outstanding mechanical properties including those of alumina based ceramics with superplastic behaviour, unconventional electromagnetic and photonic characteristics. Fabrication of metallo-dielectric highly textured composites showing electrochemical and catalytic properties is also reported. Applications of DSEC are envisaged in the fields of reinforcing fibres, wear resistant ceramics, abrasives, textured cermets and photonics. Some of these materials have been produced at industrial scale.

1. INTRODUCTION

Eutectics show unusual properties very different from those expected from the simple addition of the component phases[1]. Directionally solidified eutectics show aligned structures with single crystal domains growing perpendicular to the solidification front and connected by clean interfaces at atomic scale. The size of the phases in these materials ranges from hundreds of micrometers to tens of nanometers[2]. Moreover, unidirectional alignment of the eutectic phases along the solidification direction induces anisotropic properties in otherwise isotropic composites which can be useful in some applications. Interestingly, the size and morphology of the microstructure can habitually be modified by changing the solidification rate. The essential relationship between interlamellar spacing λ and the solidification rate v being (Hunt and Jackson law):

$$\lambda^2 \cdot v = K_I \qquad (1)$$

where K_I is a constant which depends on the eutectic system under study. In this manner, rapid solidification gives materials with very fine microstructure at the nm scale and enhanced mechanical properties[3].

In contrast to metallic eutectics, directionally solidified eutectic ceramics, DSEC, received less attention in the past due in part to difficulties in preparation because of their high melting point and high reactivity of the melts. Improvement in preparation techniques such as those related to laser melting technologies, has revitalized the study of ceramic eutectics seeking for the excellent mechanical response and thermal and chemical stability of these composites. The small phase size and the high quality of interfaces hinder the presence of large defects and as a result, DSEC showed improved mechanical strength. Moreover, the extraordinary regularity of the eutectic microstructure and the neatness of the interfaces usually prevent particle coarsening at high temperatures. The high resistance of the DSEC microstructure to homogeneous coarsening provides thermal stability to these compounds[4]. DSE oxides also present excellent resistance to oxidation owing to the intrinsic stability of these compounds and also to the absence of impurities at the interfaces[1,5]. In summary, DSEC present higher mechanical, thermal shock resistance and fracture strength than single crystals and

185

glasses, and a better thermal stability and retention of mechanical resistance up to temperatures near to the melting point than conventional ceramic[6].

Because of the large diversity of microstructure morphologies that can be obtained in ceramic eutectics, current interest for DSEC also expands to functional applications. Recently, some different applications namely in the fields of photonics, electronics and magnetism have been foreseen: In particular the possibility of creating dielectric-dielectric, dielectric-semiconducting (or metallic), dielectric-ferroelectric, eutectics with aligned microstructures could potentially generate a new category of composite functional materials.

2. FABRICATION METHODS

Many of the foreseen applications for ceramic eutectics require a good resistance to sliding/rolling wear, erosion, oxidation, and corrosion, as well as good chemical stability and long-term durability and high strength from cryogenic to elevated. Furthermore it is necessary to develop reliable and economical manufacturing methods. On the other hand, to obtain the desired fine and homogeneous microstructure we need step solidification thermal gradients. High thermal stresses developed during directional solidification promote a limitation to the ceramic piece size[1].

In the beginning of the DSEC studies well established bulk crystal growth techniques like Czochralski or Bridgman growth were used. They can produce relatively large DSEC pieces but the solidification rates in these techniques are low and consequently, the microstructure of the materials grown using these methods is coarse[7]. Nowadays, the so-called shaped growth techniques offer remarkable advantages over these conventional methods, the most important being the capability for production of ready-to-use samples such as ribbons, tubes or fibers and the structural and chemical perfection obtained by these new methods. In fact, the best mechanical behavior is found in fibers grown that way. Of great significance is the float zoning method because of the absence of any crucible which permits the melt of high temperature ceramics[8,9]. Zirconia was one of the components of several eutectics grown using these techniques which also involved aluminum, magnesium, calcium and yttrium oxides. These DSEC were grown in the form of fibers of diameter ranging from a few millimeters to hundreds of microns in diameter. However, the growth aspects leading to structural and shape perfectness are still far from being completely understood.

2.1. Laser zone techniques

The method uses an original optical system to focus the laser onto the fiber in an axially symmetric irradiance field (reflaxicon). The melt is self-sustained by the liquid surface tension avoiding the need of melt containers. This method is also called the Laser Floating Zone method (LFZ). The basics of the technique is the formation of a small melt liquid zone that moves at a controlled speed throughout the precursor ceramics generating a directionally solidified rod. The major advantages of the float zone techniques are the absence of contamination from crucible materials, the high melt temperatures achievable and the possibility of easy control of the solidification atmosphere. One of the main restrictions of the technique is the geometry of the samples; they are cylinders with a limited diameter that depends on the mechanical and thermal properties of the materials and on the thermal gradients at the interface of solidification and heating power.

Other heating sources on zone melting techniques employ RF induction coupling[10], electron beam[11], or thermal imaging (mirror furnace)[12] heating. The so called immersed-heater process or modified floating zone technique uses a perforated and heated iridium or platinum strip immersed in the melt during the growth process.

CO$_2$ laser
(P$_{max}$ = 600 W)

Rotation motor

Monitor for
viewing fibers

Upper specimen
holder

Computer

Figure 1. LFZ equipment set-up.

Figure 1 shows a basic laser floating zone growth system consisting of a CO$_2$ or Nd:YAG laser as heating source, a growth chamber with an optical system for the laser beam focusing and two vertical axes for the ceramic cylinder and solidified rod displacement. The laser beam inside the chamber is usually deflected by a reflaxicon that transforms the circular laser spot into a ring-shaped beam (Figure 2). The growth process starts by heating the end of the precursor until a melt drop is formed. A seed placed in the other axis is brought into contact until a liquid bridge between the precursor and seed is established. During the growth, the cylinder precursor is pushed towards the liquid zone and the seed withdrawn from it. The apparatus permits the simultaneous opposite sense rotation during growth of both the eutectic rod and the feed material rod. When the fiber is grown upward from a sintered powder rod, some gas bubbles can be trapped within the grown crystal. The downward growth eliminates most of the bubbles but at high growth rates they can still persist inside the crystal. Because the axes are moved by step or continuous motors, vibrations can be generated due to the friction between these and the elements of the chamber, provoking small perturbations during growth that can be reflected in defects affecting the quality of the resulting samples. Other commonly found instabilities come from laser power fluctuations, non uniformity in heat distribution, poor alignment of the rods, excessive zone length, etc. The maximum length of the large fiber depends on the limitations of the movement systems.

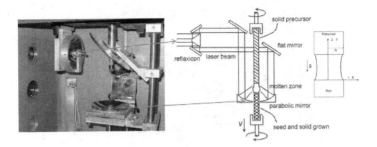

Figure 2. LFZ growth chamber

2.2. Laser surface melting technique

Laser surface melting is a recently developed technique to produce flat plates of directionally solidified eutectics[13,14]. In Figure 3 we give a diagram of the surface directionally solidifying process. The surface of the material can be covered by overlapping single CO_2 laser beam tracks or by a laser spot shaped as a line. The laser line scans over the precursor surface producing a moving melt pool. The depth of the melt pool depends on several parameters such as material properties, laser power and processing speed. Usual travelling speeds in surface processing are in the range 20–1500 mm/h and the ceramic can be melted to depths from 50 to 1500 μm.

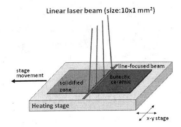

Figure 3. Scheme of a laser surface melting set-up

In figure 4 we show some of the materials processed by this method

Figure 4. Al$_2$O$_3$/YSZ eutectic ceramic plates before and after laser surface melt processing.

2.3. Edge-defined, film-fed growth (EFG)

In 1959, A.V. Stepanov[15] produced shaped crystals by growing from melt using a molybdenum or tungsten shaper. The technique allows direct pulling of the crystals with a variety of cross-sectional shapes such as tubes or ribbons. The EFG process is a particular case of the Stepanov method, that uses wettable shapers. A liquid pool is formed on the planar top surface of the shaper which feeds from a liquid reservoir by capillary through the die from which the crystal is withdrawn. In the EFG method it is the shape of the die and not the melt column, which controls the shape of the crystal. Finch et al.[16] grew the Mn$_2$SiO$_4$/MnO eutectic using the EFG technique by the first time. Startostin et al.[17] described the microstructure and crystallographic textures of phases in crystals of Al$_2$O$_3$/ZrO$_2$(Y$_2$O$_3$) directionally solidified using the EFG technique.

The μ-PD technique is a well established technique derived from EFG for fiber growth which makes use of a die placed at the bottom of the melt crucible from which the fiber is withdrawn. The small convection currents and meniscus at the bottom yields to a precise and stable control of the growth front and to a homogeneous microstructure throughout the entire sample cross-section. The method also has the advantage of making unnecessary the preparation of source samples. The first development of single crystalline fibre growth by the μ-PD method was established at the Fukuda's Laboratory in Japan. The method has been thoroughly described by Yoon and Fukuda[18].

3. CERAMIC EUTECTIC SYSTEMS

A large diversity of microstructure morphologies can be found in ceramic eutectics. Geometrical motifs such as lamellar, fibrous, complex-faceted, sometimes called chinese-script microstructure, etc., have been reported to occur in various DSEC's (see Figure 5). This morphological diversity is in part due to the high melting entropy values of the crystalline phases which make it difficult to create the necessary conditions for regular growth. Reference to earliest work on the microstructure and crystallography of ceramic eutectics can be found in the reviews of Ashbrook[19],

Stubican and Bradt[7] and Revcolevschi et al.[20].

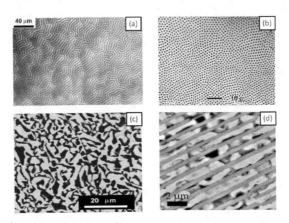

Figure 5. Micrographs showing the microstructure of some DSEC a) Transmission optical microscope image of the CaSZ(brighte)/CaZrO$_3$ eutectic showing a meander-type microstructure, b) SEM image of a fibrous MgO(rods)/MgSZ eutectic, c) Chinese script microstructure of a Al$_2$O$_3$/YAG(bright) DSEC, d) lamellar structure of a Co/YSZ cermet

In recent times, attention has been paid to oxide ternary compounds searching for a phase refinement[2]. Perfectness in microstructure makes fibrous eutectics the easiest way of obtaining good quality, single crystal fibers of ceramic compounds. In fact LiF single crystals less than 1 μm diameter and of millimeter length range were obtained from a NaCl/LiF eutectic after removing the NaCl matrix[21]. MgO single crystal fibers are also found in the MgF$_2$/MgO and CaF$_2$/MgO eutectics. In the latter light is transmitted by the highest refractive index MgO micron size single crystalline fibers[22]. Two different eutectic compositions can be found in the same compound, as is the case of the Pr$_2$O$_3$/Al$_2$O$_3$ eutectic. The PrAlO$_3$/Pr$_2$O$_3$[23] and the PrAlO$_3$/PrAl$_{11}$O$_{18}$**Error! Bookmark not defined.** DSEC show interesting luminescence properties and the latter gives the first evidence of the existence of crystalline PrAl$_{11}$O$_{18}$. For the sake of completeness we want to mention the special case of CaSiO$_3$/Ca$_3$(PO$_4$)$_2$ eutectic. Eutectic compositions can also promote glass formation, as in the above mentioned case where a eutectic glass with excellent optical properties can be produced[24].

4. APPLICATIONS OF EUTECTIC CERAMICS
4.1. Mechanical properties. Structural Applications
The working temperature for materials in systems for energy generation is continually increasing to the point where new materials others than nickel and cobalt–based superalloys must be developed. Pure oxide ceramics present high stability in air at high temperatures but they show low tensile strength and creep resistance. Additionally, ceramic components contain many grain boundaries that are not thermodynamically stable at high temperatures. Directionally solidified eutectics based on Al$_2$O$_3$ (i.e. Al$_2$O$_3$/Y$_2$O$_3$ or Al$_2$O$_3$/ZrO$_2$(Y$_2$O$_3$), Al$_2$O$_3$/Gd$_2$O$_3$ systems) improve on the mechanical properties of most ceramics because of their nearly perfect component arrangements (fibrous, lamellar or no ordered homogeneous interpenetrating phases).

4.1.1. Mechanical strength

Among the oxide eutectics, the directionally solidified materials of the Al_2O_3-ZrO_2-Y_2O_3 system have been confirmed to be the best in terms of their mechanical behavior. The sapphire-yttrium garnet eutectic ceramic (Al_2O_3/YAG) shows exceptional mechanical resistance in air up to temperatures above 2000 K, but with relatively low toughness values ≈ 2 MPa$\sqrt{m}$. The flexural strength of Al_2O_3-YAG eutectic was measured by Waku et al.[25] as a function of temperature and compared with that of a sintered composite of the same composition. The Al_2O_3-YAG eutectic maintained its room temperature strength in the range of 350-400 MPa up to 1880 °C, just below melting point. A systematic study of the mechanical properties of several eutectics based on mixtures of Al_2O_3 and rare-earth oxides was reported by Yoshikawa[26]. Flexural strength values up to 1.5 GPa and toughness of 7,8 MPa m$^{1/2}$ were reported in Al_2O_3-ZrO_2 (Y_2O_3) eutectics at room temperature by Pastor et al.[27]. Similar data were reported by Borodin et al.[28]. Courtright et al.[29] obtained higher values of mechanical resistance, between 1.13 and 2.45 GPa and verified that samples treated up to 1560°C for 24 hours did not present important losses of their mechanical properties. Hulse and Batt[30] observed in bending tests by three points, plastic deformation at very high temperature and obtained flexural strength values of 524 MPa at 1575°C, among the highest published to date for oxides at these temperatures. In table 1 the values of some physical properties of different eutectics are compared with those of some ceramic materials. Notice the outstanding mechanical resistance of some ceramic eutectics and moderate toughness values in most of them.

Table 1. Flexural strength, σ_f, and toughness, K_c, of some eutectic ceramics compared with those of some industrial ceramics.

Material	σ_f (GPa) at 300 K	σ_f (GPa) (at T indicated)	K_c (MPam$^{1/2}$)
YSZ/ Al_2O_3	1.55	0.9 (1600 K)	5-7.8
Al_2O_3/$Y_3Al_5O_{12}$	1.9	1.7 (1700 K)	2
Al_2O_3/$Y_3Al_5O_{12}$/YSZ	4.0	2.5 (1700 K)	3.3
Al_2O_3/$Er_3Al_5O_{12}$	2.7	2.7 (1600 K)	1.9
YSZ (ceram.)	0.24		5
SiC (ceram.)	0.21-0.38		4
Al_2O_3 (ceram.)	0.33		4

4.1.2 Creep

Al_2O_3/YAG and Al_2O_3/ZrO_2(Y_2O_3) eutectic oxides directionally solidified with microstructure size above 1 μm have a remarkable creep resistance because the strong interfaces prevent interfacial sliding. Sayir et al.[31] studied creep resistance of Al_2O_3/ZrO_2(Y_2O_3) eutectic at 1400 °C and 300 MPa, concluding that this eutectic is an order of magnitude more resistant than the single crystalline cubic zirconia and even superior than the off-axis sapphire. They also verified that the resistance of these eutectics remains almost constant up to 1400°C, contrasting with sapphire single crystals that presented a sudden deterioration. Waku et al. [32] observed that plastic deformation mechanism in Al_2O_3-YAG eutectics is controlled by dislocation motion. However, in compounds with finer microstructure mass transport mechanisms (diffusion mechanisms) can also play an important role on creep.[33]

The ternary compound Al_2O_3-ZrO_2-YAG when produced by LFZ at 1200 mm/h shows a superplastic behaviour at 1700 K with strain rates as high as 1mm/min. This material exhibits an outstanding flexural strength of about 5 GPa at room temperature, the highest strength ever reported in a bulk material[2]. These exceptional properties are induced by the nonconventional microstructure formed by bundles of single-crystal c-oriented Al_2O_3 and $Y_3Al_5O_{12}$ whiskers of ~100 nm width and with smaller YSZ whiskers between them

4.1.3. Wear

There have only been a few studies on the wear resistance of DSEC in bulk or coating form. Miyoshia et al.[34] subjected near-eutectic Al_2O_3/ZrO_2(Y_2O_3) rods grown by LFZ to reciprocating sliding friction experiments against B_4C plates and showed satisfactory wear resistance in the 296–1073 K temperature range. Similar high values of the specific wear resistances were reported by Ester el al.[35] in fully dense, homogeneous, and crack-free Al_2O_3/Zr_2O_3(Y_2O_3) eutectic oxide coatings obtained by laser melting of surfaces of sintered ceramic plates. The Vickers hardness, friction coefficient, and specific wear resistance, measured at room temperature, were dependent on the solidification rate, obtaining the maximum hardness and minimum friction coefficient and wear rate in samples processed at 500 mm/h. It was almost an order of magnitude smaller than that of the conventionally sintered ceramic of the same composition.

At present, the most important application at industrial scale of eutectic ceramics is perhaps that in the abrasive industry. Production of eutectic ceramic grains using fusion technologies such as arc furnaces (Herault furnace) is being performed in the Saint-Gobain Ceramic Materials Division. This technique allows the production of huge amounts of dense ceramic material for their use in abrasives, refractories, plasma coatings and recently also in SOFCs (Solid Oxide Fuel Cells).

4.2. Functional applications

DSE ceramics are also attractive as functional materials. Their multi-component character permits the fabrication of materials with mixed properties such as ferromagnetic and insulator, conductor and insulator, or electric and magnetic as was the case of the $BaTiO_3$/$CoFe_2O_4$ eutectic studied by Echigoya et al.[36]. Moreover, directional solidification induces phase alignment and hence anisotropic properties in the material. The structure of periodic and ordered arrays of alternating lamellae or fibres with sharp and clean interfaces, conveys interesting functional properties to eutectics, such as the directional transport of light and electricity among others[37,38]. The application of DSEC as substrates for deposition of textured high Tc superconductors and manganites thin films has recently been reviewed by Llorca and Orera[1]. Other applications include anisotropic electroceramics and highly textured biomaterials.

4.2.1 Textured Cermets

DSEC based on stabilized zirconia and ceria are of great significance because being good oxygen ion conductors, the stabilized zirconia (SZ) and doped ceria (GDC) phases perform as pipelines for oxygen ions moving in and out of the eutectic matrix. Oxidation and reduction of bulk ceramic eutectics is in this way feasible and opens the door to the production of a large variety of new oxide-metal composites. In this way DSEC oxides can be used as precursor materials to obtain cermets, which may find applications in the fields of heterogeneous catalysis and solid oxide fuel cell (SOFC) technology. In fact reduction of the eutectic ceramic leads to channelled Ni(Co)–YSZ (GDC) porous cermets with alternating porous metal and ionic conductor lamellae, a configuration that offers good electrical conductivity, gas permeation and a thermal expansion coefficient suitable for convenient thermo-mechanical integration with other components of the SOFC[39]. These cermets are unique in the sense that the catalyser particles, of submicron size, are confined between the YSZ fine lamellae[40] Which induces great stability against metallic particle coarsening. The effect is strongly dependent on the interface properties and the microstructure, orientation relationships, growth habits and interfaces in Ni/YSZ cermets have been recently reported[41].

4.2.2. Photonic Materials and Metamaterials

Some efforts in the study of the optical properties of DSEC have been recently made. For example, the planar optical waveguide effect was demonstrated in CaSZ/CaZrO3 lamellar regular eutectics[42] and micrometer size MgO single crystals in CaF_2/MgO DSEC perform as a bundle of optical fibers with a density of 40000 fibers/mm[22]. Another positive aspect of DSEC as photonic materials is related to the self assembled nature of eutectic structures where clusters of planar or fiber waveguides of micrometric or nanometric size are grown "in situ". In addition, most DSEC can be easily doped with optically active ions such as 3d or 4f-ions in just one or in several component phases creating an optical material with the unusual characteristic of being at the same time a monolith but also with optical properties of a multiphase material where optical active ions are placed in a number of different crystal field environments.

DSEC have also been postulated as potential metamaterials. Metamaterials are defined as artificial materials, not found in nature, with a microstructure smaller than the wavelength of the electromagnetic field showing unconventional electromagnetic properties such as negative refraction index materials, cloaking effects, giant dielectric constant, etc.[43]. Metamaterials are structured materials comprising metals and dielectrics, polymers, or magnetic oxides, which are currently fabricated by electron lithography, focused ion beam milling and other nanotechnologies. It must also be borne in mind that there is a need for new low cost fabrication technologies. The positive aspect is that eutectics are self-organized highly textured materials where realization of microstructures with geometries suitable for metamaterials fabrication is possible. For example existence of split-ring resonator-like (SSR) cross section features has been confirmed in the SrO-TiO$_2$ DSEC[44]. Moreover, the SRR structure is the first demonstrated example of a negative refraction index metamaterials[45].

5. CONCLUSIONS AND FINAL REMARKS

The mechanical and functional properties of DSEC are mainly determined by the characteristics of their dense and homogeneous microstructure. The important point here is that in DSEC materials the microstructure can be tailored by processing. New processing techniques have even opened the door to the fabrication of compositionally graded ceramics from melt as has been recently reported[46].

Deeper understanding of the correlation between processing, microstructure and properties has led to outstanding mechanical achievements, such as impressive mechanical resistance values and creep resistance as have been described in section 4. The critical point still concerns the mediocre toughness values of DSEC.

Although still in its very early stages, the study of DSE ceramics for functional applications has to receive more support in the near future. The tendency to develop micro devices perfectly matches the possibility of fabrication of relatively small pieces of eutectic ceramics with a very regular and ordered microstructure. Being an "in situ" self-assembled method, DSEC fabrication may compete with other nano and micro technologies in perfectness, cleanness and low cost.

ACKNOWLEDGEMENTS

We thank financial support by the Spanish Government under grants MAT2009-14324-C02-01 and MAT2009-13979-C03-03. The European project ENSEMBLE NMP4-2008-213669 is also acknowledged for funding

REFERENCES

[1] J. Llorca and V.M. Orera, Directionally Solidified Eutectic Ceramic Oxides, *Prog. Mat. Sci.*, **51** 711-809 (2006).

[2] P.B. Oliete, J.I. Peña, A. Larrea, V.M. Orera, J. Llorca, J.Y. Pastor, A. Martín, J. Segurado, "Ultra-High-Strength Nanofibrillar Al_2O_3-YAG-YSZ Eutectics," *Adv. Mater.*, **19** 2313-2318 (2007)

[3] H. Su, J. Zhang, Ch. Cui, L. Lin and H. Fu, "Rapid solidification behaviour of $Al_2O_3/Y_3Al_5O_{12}$ binary eutectic in situ composites," *Mat. Sci. Eng. A*, **479**, 380-388 (2008)

[4] A. Sayir, S.C. Farmer, P.O. Dickerson, A.M. Yun, "High temperature mechanical properties of $Al_2O_3/ZrO_2(Y_2O_3)$ fibers," in *Mat. Res. Soc. Symp. Proc.*, **365**, 21-27 (1993)

[5] C.O. Hulse, J.A. Batt, Final Tech. Rept. UARL-N910803-10, NTIS AD-781995/6GA. 1974

[6] Y. Waku, N. Nakagawa, T. Wakamoto, H. Ohtsubo, K. Shimizu, Y. Kohtoku, "A ductile ceramic eutectic composite with high strength at 1873 K," *Nature*, **389** 49-52 (1997)

[7] V.S. Stubican and R.C. Bradt,"Eutectic solidification in ceramic eutectics," *Ann. Rev. Mater. Sci.*, **11**, 267-297 (1981)

[8] P.B. Oliete, and J.I. Peña,"Study of the gas inclusions in $Al_2O_3/Y_3Al_5O_{12}$ and $Al_2O_3/Y_3Al_5O_{12}/ZrO_2$ eutectic fibres grown by laser floating zone," *J. Cryst. Growth*, **304**, 514-519 (2007)

[9] J.I. Peña, M. Larsson, R.I. Merino, I. de Francisco, V.M. Orera, J. Llorca, J.Y. Pastor, A. Martín and J. Segurado, "Processing, microstructure and mechanical properties of directionally-solidified Al_2O_3-$Y_3Al_5O_{12}$-ZrO_2 ternary eutectics," *J. Eur. Ceram. Soc.*, **26**, 3113-3121 (2006)

[10] M. Kimura, H. Arai, T. Mori and H. Yamagishi, "Facet formation in silicon single crystal grown by VMFZ method," *J. Cryst. Growth*, **128**, 282-287 (1993)

[11] V.N. Semenov, B.B. Straumal, V.G. Glebovsky and W. Gust, "Preparation of Fe-Si single crystal and bicrystal for diffusion experiments by the electron-beam floating zone technique," *J. Cryst. Growth*, **151**, 180-186 (1995).

[12] T. Yamakawa, N. Ishizawa, K. Uematsu, N. Mizutani, M. Kato. " Growth of yttria partially and fully stabilized zirconia crystals by xenon arc image floating zone method," *J. Crystal Growth*, **75**, 623-9 (1986)

[13] A. Larrea, G.F. de la Fuente, R.I. Merino and V.M. Orera, "ZrO_2-Al_2O_3 eutectic plates produced by laser zone melting," *J. Eur. Ceram. Soc.*, **22**, 191-198 (2002)

[14] A. Larrea, V.M. Orera, R.I. Merino, J.I. Peña, "Microstructure and mechanical properties of Al_2O_3-YSZ and Al_2O_3-YAG directionally solidified eutectic plates," *J. Eur. Ceram. Soc.* **25** 1419-1429 (2005)

[15] A. V. Stepanov, "New method of producing articles (sheets, tubes, rods, various sections, etc) directly from liquid metal.1.," *Soviet Physics-Technical Physics*, 4, 339-348 (1959)

[16] C.B Finch, J.D. Holder, G.W.Clark and H.L.Yakel, "Edge-defined, film-fed growth of Mn_2SiO_4-MnO eutectic composites. Effect of die-top geometry on solidification interface shape," *J. Cryst. Growth,* **37**, 245-252 (1977)

[17] M.Yu. Starostin, B.A. Gnesin and T.N. Yalovets, "Microstructure and crystallographic textures of the alumina-zirconia eutectics", *J. Cryst. Growth*, **171**, 119-124 (1997)

[18] Dae-Ho Yoon and T. Fukuda, "Characterization of $LiNbO_3$ micro single crystals by the micro-pulling-down method," *J. Cryst. Growth*, **144**, 201-206 (1994)

[19] R.L. Ashbrook, "Directionally solidified ceramic eutectics," *J. Am. Ceram. Soc.*, **60** 428-435 (1977)

[20] A. Revcolevschi, G. Dhalenne and D. Michel, "Interfaces in directionally solidified oxide-oxide eutectics," *Mat. Sci. Forum,* **29** 173-198 (1988)

[21] V.M. Orera and A. Larrea,"NaCl-asisted growth of micrometer-wide long single crystalline fluoride fibres," *Opt. Mat.* **27**, 1726-1729 (2005)

[22] A. Larrea, L. Contreras, R.I. Merino, J. Llorca and V.M. Orera, "Microstructure and physical properties of CaF_2-MgO eutectics produced by the Bridgman method," *J. Mater. Res.* **15**, 1314-1319 (2000)

[23] D.A. Pawlak, , K. Kolodziejak, R. Diduszko, K. Rozniatowski, , M. Kaczkan, M. Malinowski, J. Kisielewski and T. Lukasiewicz, "The $PrAlO_3$-Pr_2O_3 eutectic, its microstructure, instability, and luminescence properties," *Chem. Mat.*, **19**, 2195-2202 (2007)

[24] J.A. Pardo, J.I. Peña, R.I. Merino, R. Cases, A. Larrea and V.M. Orera, "Spectroscopic properties of Er^{3+} and Nd^{3+} doped glasses with the $0.8CaSiO_3$-$0.2Ca_3(PO_4)_2$ eutectic composition," *J. Non-Cryst. Solids*, **298**, 23-31(2002)

[25] Y. Waku, N. Nakagawa, T. Wakamoto, H. Otsubo, K. Shimizu and Y. Kohtoku, "High temperature strength and thermal stability of unidirectionally solidified Al_2O_3/YAG eutectic composite," *J. Mater. Sci.*, **33**, 1217-1224 (1998)

[26] A. Yoshikawa, K. Hasegawa, J.H. Lee, S.D. Durbin, B.M. Epelbaum, D.H. Yoon, T. Fukuda, Y. Waku, "Phase identification of Al_2O_3/$RE_3Al_5O_{12}$ and Al_2O_3/$REAlO_3$ (RE = Sm-Lu, Y) eutectics," *J. Cryst. Growth*, **218**, 67-73 (2000).

[27] J.Y. Pastor, P. Poza, J. Llorca, J.I. Peña, R.I. Merino, V.M. Orera, "Mechanical properties of directionally solidified Al_2O_3-$ZrO_2(Y_2O_3)$ eutectics," *Mat. Sci. Eng.* A, **308**, 241-249 (2001)

[28] V.A. Borodin, M.Yu. Starostin and T.N. Yalovets. "Structure and related properties of shaped eutectic Al_2O_3-$ZrO_2(Y_2O_3)$ composites," *J. Cryst. Growth*, **104** (1) 148-153 (1990).

[29] E. L. Courtrigth, J.S. Haggerty and J. Sigalovsky. "Controlling microstructures in $ZrO_2(Y_2O_3)$-Al_2O_3 eutectic fibers," *Ceramic Engineering and Science Proceedings*. **14** (7-8) 671-681. (1993).

[30] C.O. Hulse y J.A. Batt. "Effect of eutectic microestructures on the mechanical properties of ceramic oxides". (Final Technical Report N910803-10, United Aircraft Research Laboratories). Government Reports Announcements. (U.S.) **74** (19) 86. (1974).

[31] A. Sayir and S.C. Farmer, "The effect of the microstructure on mechanical properties of directionally solidified Al_2O_3/ZrO_2 (Y_2O_3) eutectic," *Acta Mater.*, **48**, 4691-4697 (2000)

[32] Y. Waku and T. Sakuma, "Dislocation mechanism of deformation and strength of Al_2O_3-YAG single crystal composites at high temperatures above 1500 ºC," *J. Eur. Ceram. Soc.*, **20**, 1453-1458 (2000)

[33] J. Ramirez-Rico, A.R. Pinto-Gómez, J. Martínez-Fernández, A.R. de Arellano-López, P.B. Oliete, J.I. Peña and V.M. Orera, "High-temperature plastic behaviour of Al_2O_3-$Y_3Al_5O_{12}$ directionally solidified eutectics," *Acta Mater.*, **54**, 3107-3116 (2006).

[34] K. Miyoshia, S. C. Farmer, A. Sayir, "Wear properties of two-phase Al_2O_3/ZrO_2 (Y_2O_3) ceramics at temperatures from 296 to 1073 K," *Tribology International*, **38**, 974–986 (2005).

[35] F.J. Ester, R.I. Merino, J.Y. Pastor, A. Martín and J. Llorca, "Surface Modification of Al_2O_3-ZrO_2 (Y_2O_3) Eutectic Oxides by Laser Melting: Processing and Wear Resistance," *J. Am. Ceram. Soc.*, **91**, 3552-3559 (2008).

[36] J. Echigoya, S. Hayashi and Y. Obi, "Directional solidification and interface structure of $BaTiO_3$-$CoFe_2O_4$ eutectic," *J. Mat. Sci.*, **35**, 5587-5591 (2000)

[37] R.I. Merino, J.I. Peña, A. Larrea, G.F. de la Fuente and V.M. Orera, "Melt grown composite ceramics obtained by directional solidification: Structural and functional applications," *Recent Res. Devel. Mat. Sci.*, **4**, 1-24 (2003)

[38] V.M.Orera, R.I. Merino, J.A. Pardo, A. Larrea, J.I. Peña, C. González, P. Poza, J.Y. Pastor and J. Llorca "Microstructure and physical properties of some oxide eutectic composites processed by directional solidification," *Acta Mater.*, **48**, 4683-4689 (2000)

[39] M.A. Laguna-Bercero, A. Larrea, J.I. Peña, R.I. Merino and V.M. Orera, "Crystallography and termal stability of textured Co-YSZ cermets from eutectic precursors," *J. Eur. Ceram. Soc.*, **28**, 2325-2329 (2008)

[40] A. Larrea, M.A. Laguna-Bercero, J.I. Peña, R.I. Merino and V.M. Orera, "Orientation relationships and interfaces in Ni- and Co-YSZ Cermets prepared from directionally solidified eutectics," *Central European Journal of Physics,* **7** (2), 245-250 (2009)

[41] M.A. Laguna-Bercero and A. Larrea, "YSZ-Induced crystallographic reorientation of Ni particles in Ni-YSZ cermets," *J. Am. Ceram. Soc.,* **90**, 2954–2960 (2007)

[42] V.M. Orera, J.I. Peña, R.I. Merino, J.A. Lázaro, J.A. Vallés and M.A. Rebolledo, "Prospects of new planar optical waveguides based om eutectic microcomposites of insulating crystals: The ZrO_2-$CaZrO_3$ erbium doped system," *Appl. Phys. Lett.,* **71**, 2746-2748 (1997)

[43] N.I. Zheludev, "The road ahead for metamaterials," *Science,* **328**, 582-583 (2010)

[44] D.A. Pawlak, S. Turczynski, M. Gajc, K. Kolodziejak, R. Diduszko, K. Rozniatowski, J. Smalc and I. Vendik, " How far are we from making metamaterials by self-organisation? The microstructure of highly anisotropic particles with SSR-like geometry," *Adv. Funct. Mater.,* **20**, 1116-1124 (2010)

[45] R. Shelby, D.R. Smith, S. Schultz, "Experimental Verification of a negative Refraction Index," *Science,* **292**, 77-79 (2001)

[46] R.I. Merino, J.I. Peña, V.M. Orera "Compositionally graded YSZ/NiO composites by surface laser melting," *J. Eur. Ceram. Soc.,* **30**, 147-152 (2010)

EFFECT OF DIFFERENT FUELS ON THE MICROWAVE-ASSISTED COMBUSTION SYNTHESIS OF $Ni_{0.5}Zn_{0.5}Fe_{1.95}Sm_{0.05}O_4$ FERRITES

A. C. F. M. Costa, D. A. Vieira, V. C. Diniz
H. L. Lira
Federal University of Campina Grande,
Department of Materials Engineering, 58970-
000 Campina Grande, PB, Brazil

D. R. Cornejo
USP – Institute of Physics
05508 - 900 São Paulo, SP, Brazil

R. H. G. A. Kiminami
Federal University of São Carlos
13565-905 São Carlos, SP, Brazil

ABSTRACT

This work involved the preparation and characterization of $Ni_{0.5}Zn_{0.5}Fe_{1.95}Sm_{0.05}O_4$ ferrite powders synthesized by microwave-assisted combustion reaction, using urea, glycine and a urea-glycine mixture as fuels. The resulting powders were characterized by X-ray diffraction, BET, scanning electron microscopy and magnetic measurements. The XRD results indicated that the powders obtained with urea presented low crystallinity, showing only the main peak of ferrite phase, while the powders obtained with glycine and, the urea and glycine mixture presented peaks corresponding to primary phase ferrite and traces of secondary phase. Surface area measurements indicated that the powders obtained using urea as fuel had a high surface area (71.44 m^2/g) exceeding that of the powders prepared with glycine or with the urea-glycine mixture, which showed values of 1.35 and 5.98 m^2/g, respectively. The powders synthesized with urea, glycine, and the urea-glycine mixture showed saturation magnetization values of 8.4, 53.0, and 46.0 emu/g, respectively.

INTRODUCTION

Ni-Zn ferrites have been studied by several researchers using different methods of chemical synthesis for various applications, e.g., soft magnetic devices, electromagnetic radiation absorbers, catalysts, pigments [1], etc. The physicochemical properties of these materials are sensitive to dopants in the spinel structure and knowledge about the type and quantity of these dopants is essential to obtain high quality ferrites for different technological applications [2]. Several techniques of chemical synthesis have shown good results in obtaining Ni-Zn ferrites with controlled morphology and structure, thereby improving the characteristics of these materials. However, most techniques do not allow for the large-scale production of these materials because they require long reaction times and produce low yields. Among the various techniques, combustion reaction synthesis has been used successfully to produce these materials on a laboratory scale with good control of their characteristics.

Several works describe the influence of the external heat source on the final characteristics of powders [1]. The use of microwaves as a source of heat for the combustion reaction employed in the preparation of Ni-Zn ferrites was described by Fu and Lin [3] in 2002 and by Costa et al. [1]. Both studies confirm that these materials can be obtained with a single-phase inverse spinel structure and with a low degree of particle agglomeration. Microwave energy offers some advantages in relation to conventional techniques, e.g., rapid and uniform heating, low synthesis temperature and lower costs in terms of energy and time [4]. Combustion synthesis requires fuel for auto-ignition and combustion, regardless of the material to be obtained (oxide or non-oxide) and the external heat source employed to activate the reaction.

Although the choice of the ideal fuel to prepare the powder of a specific ceramic system depends mainly on its cost, other factors such as valence, organic chain size, and availability are also important [5,6]. Furthermore, the type of fuel directly affects combustion temperature and time, which are important parameters that, in most cases, determine the final characteristics of the resulting powders. Urea $(CO(NH_2)_2)$ is the fuel most commonly used in combustion synthesis due to the advantages it offers, e.g., low reducing capacity with a total valence of 6+, small organic chain (low atomic mass), it produces small volumes of gas, is commercially available, low cost, and generates low combustion temperatures that suffice to produce the desired phase in the final products [7]. In recent years, however, glycine (NH_2CH_2COOH) has attracted great interest as a fuel for synthesizing certain ceramic systems by combustion reaction. This fuel, which is more expensive than urea, is an amino acid with an organic chain and high molecular weight. It is a complexing agent able to bind easily with metallic ions, and has a valence of 9+ (high reducing capacity), generating large amounts of combustion gas, and hence, high combustion temperatures [8-11].

Extensive research has focused on investigating the effect of cation doping on the spinel ferrite lattice, since it is well known that some of the properties of ferrites are sensitive to the presence of doped cations. The addition of small amounts of these doped cations changes the electrical, structural and magnetic properties of ferrites [12-13]. Various researchers have analyzed how doping of the spinel lattice of some ferrites alters their chemical composition, and hence, their intrinsic properties of magnetization, magnetic permeability, anisotropy, Curie temperature, magnetic resonance and electrical resistivity. Such doping can also change the powders' characteristics and, therefore, the microstructure formed after sintering, directly affecting the extrinsic properties of these materials. The use of samarium and its effect on the magnetic properties of Ni-Zn ferrites has been investigated by several researchers. For example, Sattar et al. [14] examined and reported on the effect of rare earth ion doping on the magnetic properties of Cu-Zn ferrites with a nominal composition of $Cu_{0.5}Zn_{0.5}Fe_{2-x}R_xO_4$ (R = La, Nd, Sm, Gd and Dy), where x = 0.0 and 0.1 in mol, prepared by the conventional oxide mixing method. Their samarium-doped samples showed a lower relative density (92.4%) than samples doped with other rare earth ions. Rezlescu et al. [15] observed similar behavior in a study of samarium-doped ferrites, attributing it to the formation of SmO_2 during the sintering process, which favored grain growth. The authors found that doped samples, especially those doped with samarium (Sm^{3+}), showed promising technological results, with 60% higher relative permeability and greater saturation magnetization than non-doped samples. Costa et al. [16] also reported on the effect of samarium doping in Ni-Zn ferrite nanopowders with a nominal composition of $Ni_{0.5}Zn_{0.5}Fe_{2-x}Sm_xO_4$ (x = 0.0, 0.05, and 0.1 mol) obtained by combustion synthesis using metallic nitrates, urea as fuel and a hot plate as the heat source. The authors reported that the nanopowders consisted of soft agglomerates of nanoparticles with a high surface area (55.8 - 64.8 m^2/g), smaller particles (18 – 20 nm) and nanocrystallite size particles. The addition of samarium resulted in a reduction of all the magnetic parameters they evaluated, namely saturation magnetization (24 – 40 emu/g), remanent magnetization (2.2 – 3.5 emu/g) and coercive force (99.3 – 83.3 Oe).

This paper describes the preparation of $Ni_{0.5}Zn_{0.5}Fe_{1.95}Sm_{0.05}O_4$ ferrite by microwave-assisted combustion reaction and discusses the influence of the type of fuel on the structural, morphological and magnetic characteristics of these powders destined for application as a soft magnetic material.

EXPERIMENTAL

$Ni_{0.5}Zn_{0.5}Fe_{1.95}Sm_{0.05}O_4$ ferrite samples were synthesized by combustion reaction using microwave energy as the heat source and $Ni(NO_3)_2.6H_2O$, $Zn(NO_3)_2.6H_2O$, $Fe(NO_3)_3.9H_2O$ and $Sm(NO_3)_3.9H_2O$ as precursor reagents (oxidizers) and cation sources. Urea ($CO(NH_2)_2$), glycine (NH_2CH_2COOH), and a mixture (1:1) of the two fuels were used as reducing agents. All the reagents were of 98% purity. Three samples were obtained according to the type of fuel used for the synthesis: 1 – urea (PU), 2 – glycine (PG), and 3 – the 1:1 wt/wt mixture of urea and glycine (PUG).

The stoichiometry employed in the synthesis of three samples followed the chemical concept of propellants and explosives [17] and consisted of a redox mixture of metal nitrates (source of cations and oxidizing agents) and fuel (reducing agent). The mixture was placed in a vitreous silica crucible and heated over a spiral coil (at close to 600°C) until it started to become more viscous and to release gas. The mixture was then transferred immediately to a microwave oven (Eletrolux model ME27S) operating at a maximum power of 950 W (100%), where it was exposed to combustion for 10 minutes.

The resulting samples were characterized by X-ray diffraction (XRD) in a Shimadzu XRD-6000 diffractometer operating with Cu Kα radiation. The average crystallite size was calculated by Scherrer's equation [23] from the reflection of the spinel structure on the (d_{111}); (d_{220}); (d_{311}); (d_{222}); (d_{400}); (d_{422}); (d_{333}) and (d_{440}) planes by means of the secondary deconvolution diffraction line of the polycrystalline silicon (used as standard). The lattice parameters were estimated using the DICVOL91 routine for Windows and FoolProof software [24], applying Vegard's Law to the cations mixture based on the X-ray diffraction patterns. The amorphous and crystalline contributions of the samples' X-ray diffraction patterns were determined by the Fourier Transform method. The resulting inverse transforms provide amorphous (IA) and crystalline (IC) diffraction intensities with eliminated background noise. The degree of crystallinity is then indicated by the ratio of these two intensities IC/(IC + IA).

The particle agglomerate size distribution was measured using a sedimentation method of the particles in the liquid phase (HORIBA CAPA-700 Particle Size Distribution Analyzer, U.S. version). The BET method (Micromeritics, Gemini – 2370 model) was employed to measure the specific surface area based on the physical adsorption of N_2 gas at cryogenic temperature. The average particle size was calculated from BET data using the equation $D_{BET} = 6/(D_t.S_{BET})$, where D_{BET} = equivalent spherical diameter (nm); D_t = theoretical density (g/cm^3) and S_{BET} = surface area (m^2/g). The particle morphology and size were examined by scanning electron microscopy (Philips XL30 FEG-SEM). Magnetization measurements were taken at room temperature (25°C) using an alternative gradient magnetometer (AGM). The values of the magnetic parameters of coercive field (Hc), remanent magnetization (Mr or σr) and saturation magnetization (Ms or σs) were determined based on these curves. The saturation magnetization was determined by adjusting the applied field data to the σ=σ (1−α/H) function, where σs is magnetization, σs is saturation magnetization, is the fitting parameter and Hc is the applied field.

RESULTS

Structural analysis

Fig. 1 shows the XRD patterns of the Ni-Zn-Sm ferrite samples obtained by microwave-assisted combustion reaction using urea, glycine and a mixture of urea:glycine (1:1) as fuels.

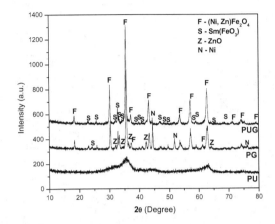

Figure 1 – XRD patterns of the Ni-Zn-Sm ferrite samples: PU - urea, PG – glycine, and PUG - urea:glycine (1:1).

As can be seen in Fig. 1, the PU sample showed a crystallinity of 43% and the formation of two peaks characteristic of inverse spinel ferrite. These peaks were broad, indicating the formation of material with small crystal sizes. These results can be explained by the specific characteristics of the fuel used for the synthesis of the sample (urea). During synthesis, this fuel releases little gas and low energy to the surrounding material, so all the heat generated in the reaction is consumed by the combustion, leading to a relatively low combustion temperature. This behavior was reported by Costa et al. [1], who studied the synthesis of different Ni-Zn ferrite systems by combustion reaction using urea as fuel and a ceramic plate (480°C) as a heat source.

The X-ray diffractogram (Fig. 1) of the PG samples showed the formation of primary phase Ni-Zn-Sm ferrite and traces of samarium and iron oxides ($SmFeO_3$), zinc oxide (ZnO) and nickel (Ni) as secondary phases. The PUG samples showed the formation of two phases: Ni-Zn-Sm ferrite phase as the primary phase, and samarium and iron oxide as secondary phases. PG and PUG presented sharp peaks for the primary phase, with crystallinity of 81% and 77%, respectively. The height of the peaks indicated the nanometric size of the particles.

The X-ray diffraction results of the Ni-Zn ferrite powder synthesized with microwaves, using glycine or the urea:glycine mixture as fuel, indicated that the combustion temperatures reached with these fuels was higher than that attained when using urea as fuel. Hwang et al. [20], who studied the effect of fuel in the synthesis of Ni-Zn ferrites on a hot plate, reported that glycine or a mixture of 1:1 urea:glycine produced a higher temperature than pure urea.

In microwave-assisted synthesis, it is impossible to verify the combustion flame temperature experimentally because the Faraday plate that protects the microwave oven does not allow the infrared pyrometer to reach the flame. Another significant effect observed when using glycine as fuel was the formation of traces of secondary phase ZnO and $SmFeO_3$. The presence of these secondary phases in the PG samples can be explained by the characteristics of glycine, namely, its highly reducing character (+9), the high molecular weight of its organic chain, and its high temperature of decomposition (≈262°C), which produces large volumes of gas and high combustion heat. The gas produced by this fuel remains in the microwave oven for a long time and the amount of oxygen is insufficient to complete the formation of the desired phase, thus probably contributing to the formation of secondary phase. The urea:glycine (1:1) mixture exhibited intermediate characteristics with regard to the phases produced in the samples. Table 1 lists the crystallite sizes, crystallinity and structural parameters of the samples synthesized by microwave-assisted combustion reaction.

Table 1 – Lattice parameters, crystallite size, and crystallinity of samples synthesized by combustion reaction

Samples	*Lattice parameters	**Crystallite size (nm)	Crystallinity (%)
PU	n.d.	38	43
PG	8.374	27	81
PUG	8.383	27	77

* Theoretical lattice parameters a = b = c = 8.399 (JCPDF-08-0234)
* *Calculated by Scherrer's equation [18]
n.d. = not determined

A comparison of the theoretical lattice parameters of Ni-Zn ferrite (8.399, JCPDF 08-0234) and the lattice parameters estimated for the samarium-doped powders revealed a discrete reduction of 0.3 and 0.2% in the PG and PUG samples, respectively. This minor reduction in the lattice parameters can be attributed mainly to the substitution of Fe^{3+} ions by Sm^{3+} in the spinel structure, since the ionic radius of Sm^{3+} (1.81 Å) is 31.5% larger than that of Fe^{3+} (1.24 Å), both with octahedral coordination.

However, in the case of the PG sample, it was found that the presence of $SmFeO_3$ and ZnO secondary phases can also contribute to this reduction in lattice parameters. In this case, Sm^{3+}, Fe^{3+} and Zn^{2+} ions segregate, creating a secondary phase that diminishes their effective contribution to the formation of primary phase. This was evidenced by comparing the estimated lattice parameters of PG and PUG with the presence of two secondary phases and the reduction in the lattice parameters of the PG sample.

The estimated crystallite size, crystallinity and lattice parameters increased significantly in the urea:glycine (1:1) mixture. This behavior was similar to the results obtained for PG sample. However, the samples prepared with the fuel mixture (PUG) showed a lower crystallinity than those prepared with glycine, which presented secondary phase $SmFeO_3$ and traces of ZnO. This finding indicates that the presence of secondary phase can contribute to increase the crystallinity of the primary phase, since the samples prepared with the fuel mixture showed the formation of secondary phase $SmFeO_3$ and of primary phase Ni-Zn-Sm.

In general, the results of crystallite size, crystallinity and lattice parameters of the PUG samples displayed a combination of the characteristics of both fuels used for PU and PG, respectively. On the other hand, the samples produced with the fuel mixture presented structural characteristics that were more similar to those of the samples produced with glycine than with urea. In an evaluation of the effect of glycine, polyethylene glycol and urea on the synthesis of pure and Mn-doped cerium, Murugan et al. [21] found that glycine resulted in the formation of samples with 25% larger crystallite sizes than samples prepared with a stoichiometric composition of mixed glycine and urea. This behavior was the opposite of that reported in this work when glycine and urea were used in the preparation of Ni-Zn-Sm ferrite, since there was a reduction of 29% when using glycine, urea and a mixture of urea:glycine. It is quite evident that the type of fuel used, or a mixture of two fuels, significantly affects the final characteristics of Ni-Zn-Sm ferrite samples.

Morphological analysis

Fig. 2 shows the values of the equivalent spherical diameter as a function of cumulative mass for the PU (Fig. 2a), PG (Fig. 2b) and PUG (Fig. 2c) samples. The values indicate the formation of soft nanoparticle agglomerates.

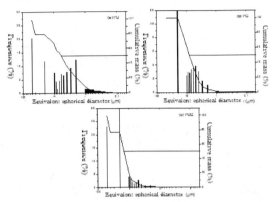

Figure 2 – Equivalent spherical diameter as a function of cumulative mass (a) PU, (b) PG and (c) PUG.

The PUG sample (Fig. 2c) showed a narrower particle size distribution than the PU and PG samples, although the difference was greater in relation to PU (Fig. 2a), which showed a large particle size distribution. With regard to the agglomerate sizes, the PUG sample showed a size of 14 µm, while the PU and PG samples showed sizes of 6 and 8µm, respectively. All the procedures resulted in agglomerates with soft characteristics. The use of glycine (PG) increased the tendency for particle agglomeration, which in turn increased the agglomerate size in comparison to the PU samples processed with urea. This was probably due to the high combustion temperature reached during synthesis.

Table 2 shows the characteristics of the PU, PG and PUG samples (surface area, particle size, and crystallite size). The PU sample showed a surface area of 71.44 m^2/g, which was 98.2% and 91.6% larger than the surface areas of the PG and PUG samples, respectively, thus confirming its small particle size.

These results indicate that the synthesis conditions strongly influence the morphological characteristics of samples. As mentioned earlier, the reaction performed with urea as fuel produced a low combustion temperature and prevented pre-sintering and/or increases in particle size when compared with the reaction using glycine and glycine:urea (PG and PUG). The results obtained from the relation between the particle size estimated by BET and the crystallite size estimated by Scherrer's equation ($T_{p(BET)}/T_{c(XRD)}$) for the samples obtained by PG and PUG were higher than those obtained by PU, indicating that the particles were polycrystalline and agglomerated. From the $T_{p(BET)}/T_{c(XRD)}$

relation it was observed that the PG sample reached a high state of agglomeration almost four-fold that of the PUG sample.

Table 2 – Characteristics of the PU, PG and PUG powders produced by microwave-assisted combustion reaction

Samples	Surface area (m^2/g)	Particle size $T_{p(BET)}$* (nm)	Crystallite size $T_{c(DRX)}$** (nm)	$T_{p(BET)}/T_{c(DRX)}$
PU	71.44	16	38	0.4
PG	1.35	839	27	31
PUG	5.98	189	27	7

*Estimated from surface area
**Estimated from X-ray diffraction data and using Scherrer's equation [18]
Dt = 5.361 g/cm3 JCPDS (52-0278).

A discrepancy was found between the crystallite size and particle size (D_{BET}) of the PU sample estimated from the nitrogen adsorption analysis. This discrepancy was attributed to the amount of amorphous material contained in the composition prepared with urea (PU). Therefore, the smaller particle size compared to the crystallite size was justified by the large amount of amorphous material in the sample, which was not taken into account when it was estimated from the X-ray diffraction data.

Fig. 3 shows the morphology of the PU, PG and PUG samples. These micrographs revealed that the PU particles were very small (on a nanometric scale), with a tendency to form soft porous agglomerates in response to weak van der Waals forces, but that are easily deagglomerated. The PG and PUG samples showed the formation of large irregular plate-shaped agglomerates. The morphology of the PG samples was more porous than that of the PUG samples. However, the PUG samples presented larger agglomerates and a narrower distribution than the PG samples. A comparison of the samples produced with urea and with glycine indicated that glycine tended to generate a higher combustion temperature, and to increase and pre-sinter the particles, tending to produce larger and harder agglomerates than those produced when using urea. The PG and PUG samples were also found to display similar characteristics, which is a strong indication that the effect of glycine predominated over that of urea.

An analysis of the micrographs of the samples obtained by combustion reaction, correlating the oxygen balance (OB) when using glycine as fuel, indicated that some of the results of this work were similar to those reported by Hwang et al. [22]. These authors reported that the quantity of gases emitted during the reaction increased the porosity and size of the agglomerates.

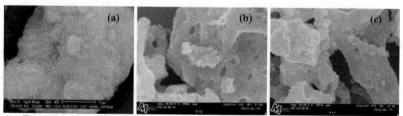

Figure 3 – SEM micrographs of the samples obtained by combustion reaction: (a) PU, (b) PG, and (c) PUG.

Magnetization analysis

The hysteresis loop in Fig. 4 shows the dependence of magnetization (M) on the applied magnetic field (H) of the $Ni_{0.5}Zn_{0.5}Fe_{1.95}Sm_{0.05}O_4$ samples. The values of the magnetic parameters, i.e., coercive field (H_C), remanent magnetization (M_r or σr) and saturation magnetization (M_s or σs), were determined by $\sigma \times H$ curves. The hysteretic loss was estimated from the measurement of the area of the hysteresis loop (W_B). These parameters are listed in Table 3.

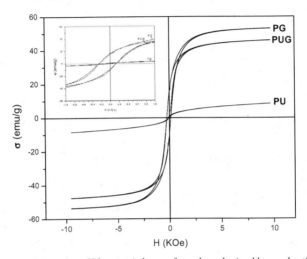

Figure 4 – σ-H hysteresis loops of powders obtained by combustion reaction.

As can be seen in Fig. 4, all the samples displayed the characteristics of a soft magnetic or permeable material (easy magnetization and demagnetization) due to the very narrow σ-H hysteresis loop. However, the type of procedure significantly influenced the area of the hysteresis curve, and a considerable reduction was observed in the coercive field and saturation magnetization as a function of the different types of fuel. A slight decrease was observed in the loop area and, hence, a decrease in hysteretic loss. The samples prepared by procedure 1 showed lower values of magnetization and coercive field than the PG and PUG samples. This was due to the nanometric size (16 nm) of the urea-synthesized particles estimated by BET when compared to the samples prepared with glycine (839nm) and with the fuel mixture (189nm). It is well known that nanometric particles have small domain size, so their boundaries or domain walls prevent rotation and/or spin movements, contributing to reduce their magnetization.

Table 3 lists the values of coercive field (H_c), remanent magnetization (M_r), saturation magnetization (M_s), squareness, and hysteretic loss measured by the integrated area of the loop. Note that the PG and PUG samples show a similar magnetic behavior, i.e., they present the same coercive field. The saturation magnetization and the integrated area (hysteretic loss) of PG were only 13% and 11% higher, respectively, than the values obtained for the PUG sample. In a study of the Ni-Zn ferrite without dopant, Costa et al. [23] found values of Ms = 58.74 emu/g and Hc = 0.88 kOe. A comparison of these values with those obtained in the present work indicated that the samples prepared with glycine and with the urea:glycine mixture using microwave energy presented, respectively, 9.7% and 21.6% lower Ms values and higher Hc values than those obtained by the aforementioned authors. This decrease in the magnetic parameters was the direct consequence of the presence of secondary phases, which allow for electronic jumping and contribute to the reduction of saturation magnetization, increase of the coercive field, and loss of hysteresis.

Table 3: Hysteresis parameters of the Ni-Zn ferrite doped samarium.

Samples	Hc (KOe)	Mr (emu/g)	Ms (emu/g)	Mr/Ms	Wb (emu/g x KOe)
PU	0.10	0.98	8.36	0.12	55.98921
PG	0.25	15.60	53.0	0.294	414.75066
PUG	0.25	15.60	46.1	0.338	367.30120

CONCLUSIONS

Microwave-assisted combustion synthesis can be recommended as a promising method. In this work, the method was employed successfully in the preparation of Ni-Zn-Sm crystalline nanoferrite, proving less time-consuming than other methods and resulting in particles with a narrow size distribution. The final characteristics of the resulting samples can vary significantly as a function of the combustion fuel. Glycine led to the formation of a highly crystalline sample and to the presence of secondary phases, unlike the samples produced with urea and with the urea:glycine mixture. The sample synthesized with urea presented an 89.84% larger surface area than the samples prepared with glycine and with the urea:glycine mixture, as well as smaller particle sizes and agglomerates. The SEM images revealed that the morphology of the samples produced by the three procedures was typical of ferrite samples obtained by combustion reaction (soft agglomerates with weak van der Waals forces). The PG and PUG samples showed good and relatively similar Ms and Hc values. The use of urea led to the formation of samples with low magnetic parameters, which resulted from the samples' highly nanometric characteristics.

ACKNOWLEDGEMENTS

The authors would like to thank to CNPq, RENAMI-CNPq, FAPESP and CAPES by support this research.

REFERENCES

[1] Costa, A. C. F. M; Morelli M. R.; Kiminami R.H.G.A., Combustion Synthesis Processing of Nanoceramics. *Handbook of Nanoceramics and Their Based Nanodevices*, 1, (2009) chapter 14, 375-391.
[2] Costa, A. C. F. M., PhD thesis, In Portuguese, Universidade Federal de São Carlos, Brazil (2001).
[3] Fu, Y.-P.; Lin, C.-H., *J. Magn. Magn. Mater.*, 251 (2002) 74–79.

[4] D. E. Clark, W. H. Sutton and D. A. Levis, *"Microwave Processing of Materials", in Microwaves: Theory and Application in Materials Processing IV*. Ceram. Trans., 80 Edited by D. E. Clark, W.H. Sutton and D. A. Levis. Amer. Ceram. Soc., OH 61-96 (1997).

[5] Costa, A. C. F. M., Lula, R. T., Kiminami, R. H.G. A., Gama, L. F. V., De Jesus, A. A., Andrade, H. M. C., *J. Mater. Sci.* 41 [15] (2006) 4871- 487.

[6] J. J. Kingsley and K. C. Patil, "A Novel Combustion Process for the Synthesis of Fine Particle α-Alumina and Related Oxide Materials", *Mater. Lett.* 6 [11,12] (1988) 427-32.

[7] Segadães, A. M.; Morelli, M. R.; Kiminami, R. H. G. A., *J. Eur. Ceram. Soc.*, 8 (1998) 771-781.

[8] Purohit, R. D.; Sharma, B. P.; Pillai, K.T.; Tyagi, A. K., *Mat. Res. Bull.*, 36 (2001) 2711.

[9] Wu, K. H.; Ting, T. H.; Li, M. C.; Ho, W. D., *J. Magn. Magn. Mater.* 298 (2006) 25–32.

[10] Lenka, R. K.; Mahata,T.; Sinha, P. K.; Tyagi, A. K., *J. Alloys Compd,* 466 (2008) 326–329.

[11] Sathi, R.; Purohit, R. D.; Tyagi, A. K.; Sinha, P.K.; Sharma, B. P. *Mat. Res. Bull.* 43 (2008) 1573–1582.

[12] Rana, M. U.; Abbas, T. *J. Magn. Magn. Mater.* 246 (2002) 110.

[13] Znidarsic, A.; Limpel, M.; Drofenik, M. *IEEE Trans. Magn.* 18 (6) (1982) 1544.

[14] Sattar, A. A.; Samy, A. M.; El-Ezza, R. S.; Eatah, A. E. *Phys. Stat. Sol.,*193 (2002) 86.

[15] Rezlescu, N.; Rezlescu, E.; Pasnicu, C.; Craus, M. L. *J. Magn. Magn. Mater.* 136 (1994) 319.

[16] Costa, A. C. F. M.; Diniz, A. P. A.; Melo, A. G. B.; Kiminami, R. H. G. A.; Cornejo, D. R.; Costa, A. A.; Gama, L., *J. Magn. Magn. Mater.*, 320 (2008) 742–749.

[17] Jain, S. R.; Adiga, K.C.; Verneker, V.P., *Combustion Flame,* 40 (1981) 71-79.

[18] Klung, H.; Alexander, L., *In X ray diffraction procedures* New York: Wiley, 1962, 25.

[19] Louer D., Roisnel T., Laboratorie de Cristallochimie, Universite de Rennes I,Campus de Beaulieu,France,1993.

[20] Hwang, C.-C.; Wu, T.-Y.; Wan, J.; Tsai J.-S., *Mat. Sci. and Eng.,* B 111(2004) 49-56.

[21] Murugan, B.; Ramaswamy, A. V.; Srinivas, D.; Gopinath, C. S.; Ramaswamy, V., *Acta Mat.*, 56 (2008) 1461–1472.

[22] Hwang C-C, Tsai J-S, Huang T-H, Peng C-H, Chen S-Y., *J. Sol. Sta. Chem.*, 178 (2005) 382–389.

[23] Costa, A. C. F. M.; Silva, V. J.; Cornejo, D. R.; Morelli, M. R.; Kiminami, R. H. G. A.; Gama, L., *J. Magn. Magn. Mater.*, 320 (2008) e370–e372.

Author Index